Civil Engineering and
Engineering Mechanics Series

N. M. Newmark and W. J. Hall, editors

INTRODUCTORY STRUCTURAL ANALYSIS

PRENTICE-HALL, Inc.
Englewood Cliffs, New Jersey

WITH
MATRIX METHODS

CHU-KIA WANG

Professor of Civil Engineering
The University of Wisconsin-Madison

Library of Congress Cataloging in Publication Data

WANG CHU-KIA, 1917–
Introductory Structural analysis with matrix methods

(Civil Engineering and Engineering Mechanics Series)
1. Structures, Theory of. 2. Matrices. I. Title
TA642. W28 624'. 171 72–667
ISBN 0–13–501650–9

10 9 8 7 6 5 4 3 2 1

Printed in the United States of America

PRENTICE-HALL, INC., *London*
PRENTICE-HALL OF AUSTRALIA, PTY. LTD., *Sydney*
PRENTICE-HALL OF CANADA LTD., *Toronto*
PRENTICE-HALL OF INDIA PRIVATE LIMITED, *New Delhi*
PRENTICE-HALL OF JAPAN, INC., *Tokyo*

Contents

Preface

This textbook has been prepared for use in the first course in structural analysis, after the statics and strength of materials courses. Because of the increasing use of computers in professional structural engineering, matrix methods have been integrated with conventional methods throughout the text. Since the objective here is to provide an understanding of matrix methods, rather than of their implementation on the computer, there is no inclusion of computer programs.

An attempt has been made to use sentences as simple, exact, and essential as possible in deriving and demonstrating the various methods of structural analysis. The reader is urged to examine and assimilate the text material critically, steadily, or even repeatedly, and to study and work through every numerical example. Exercises, either similar or supplementary to the examples, are inserted at appropriate points along the journey of study; the intention here is that every suggested exercise should be carried through in perfect detail and draftmanship by each student. In this sense, the text is designed as a workbook for programmed study. Experienced teachers may wish to delete, substitute, or add a variety of exercises with more practical connotations.

Considerable effort has been made to develop and substantiate the important energy theorems in Chapter 8. Because the two Castigliano's theorems are useful in the derivation of stiffness and flexibility matrices of finite elements in plate and shell structures, this chapter may also be useful at the beginning of a finite element course.

Mrs. Audrey M. Miller, Secretary of the Structural Engineering Faculty at the University of Wisconsin-Madison, has provided excellent assistance in typing this manuscript.

C. K. WANG
Madison, Wisconsin

Structural Analysis

1

1.1 Framed Structures vs. Plate or Shell Structures

Structures are built to withstand loads throughout the course of their use. Common structures include bridges, buildings, ships, airplanes, machineries, dams, and many other load-carrying frameworks. In almost all cases, the complete assembly may be separated into parts, each of which is again considered as a unit in its analysis and design. For a structural unit, distinction may be made between a *framed structure* and a *plate* or *shell structure*.

A framed structure is composed of straight or curved bars, whose ends may be either pinned or welded together. Typical framed structures are the trusses, the beams, and the rigid frames. An ideal two-dimensional truss as shown in Fig. 1.1a consists of bars connected by pins inserted at their ends. The continuous beam of Fig. 1.1b, although it may actually be made of one

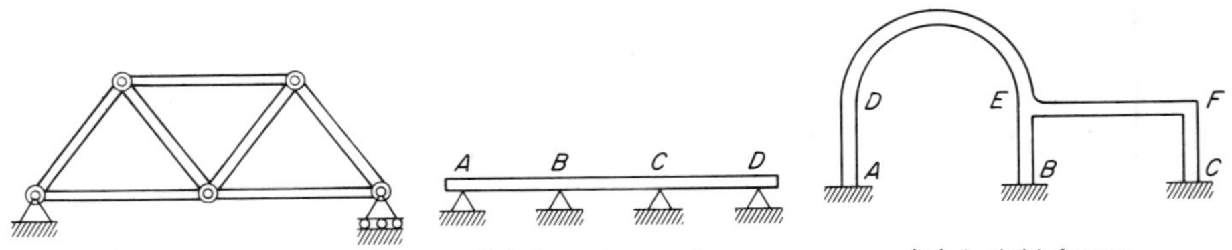

(a) A truss (b) A continuous beam (c) A rigid frame

FIG. 1.1. Typical Framed Structures.

long member, may be considered to be three bars, butt welded at *B* and *C*. A model of the two-dimensional rigid frame of Fig. 1.1c may be actually cut from a steel plate, but in its prototype it may be a monolithic reinforced concrete frame, or a steel frame of five members with welded joints at *D*, *E*, and *F*.

A plate or shell structure is built of continuous flat or curved sheets usually supported continuously at the edges. Typical examples of plate or shell structures are the floor slabs and domed roofs of buildings, deck plates and hulls of ships, wings and fuselages of airplanes, etc. Analysis of regularly-shaped plate and shell units is usually the subject matter of the Theory of Elasticity. More recently, approximate numerical analysis of irregular plates and shells has been made by the finite element method.

It is the primary purpose of this textbook to introduce the methods of analyzing framed structures; thus *Introductory Structural Analysis* is the title of this book.

1.2 First-Order Analysis vs. Second-Order Analysis

The analysis of a framed structure, whether it be a truss, a beam, or a rigid frame, involves the determination of

1) the deflected shape of the structure; and
2) the axial force, and/or the shear and bending moment at any one point in any one member on the basis of
   a) the given shape of the structure;
   b) the sizes and material properties of the members; and
   c) the applied loads.

To illustrate this, the possible deflected shapes of the three typical structures are shown in Fig. 1.2.

In the case of a truss, the deflected shape is defined by the horizontal and vertical displacements of each joint, since all bars remain straight even after loads are applied. Each joint is subjected to a system of concurrent forces which include the applied loads at this joint and the pulls on it by the bars in tension, as well as the pushes on it by the bars in compression. Now there is

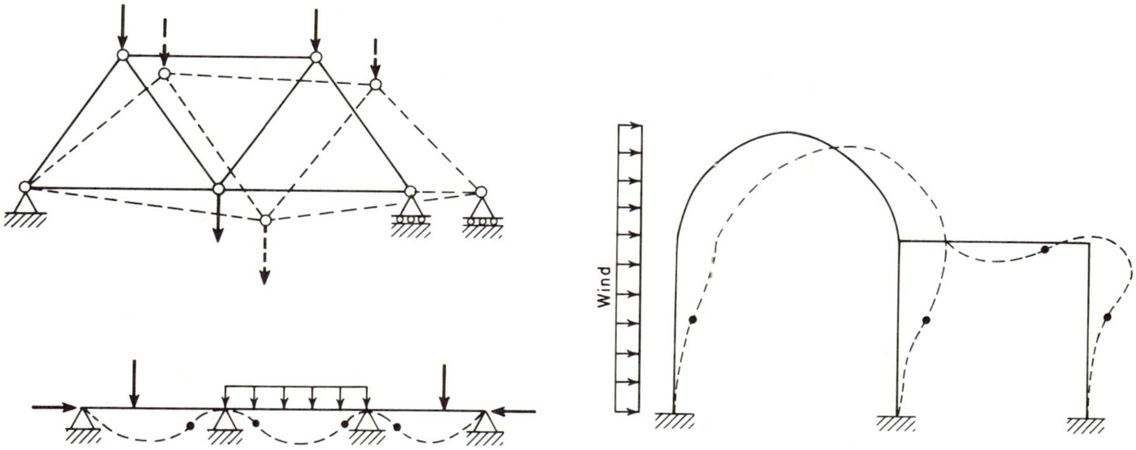

FIG. 1.2. Possible Deflected Shapes of a Truss, a Beam, and a Rigid Frame.

the question of whether the directions of the pulls and pushes should be those of the solid lines in Fig. 1.2 or those of the dashed lines in the same figure. The approach used in the former, where the equilibrium equations are based on the geometry of the original shape, belongs to *first-order analysis*; but that of the latter, where the equilibrium equations are based on the deformed geometry, is theoretically more accurate, is one further step ahead, and is therefore given the name *second-order analysis*. Second-order analysis is a complicated affair because the deformed geometry and the axial forces in the bars are mutually implicit. Fortunately, in ordinary cases the changes in geometry are so small that second-order analysis is not needed.

The elastic curve for the continuous beam in Fig. 1.2 is a function of the bending moment which is due primarily to the transverse loads on the beam. In first-order analysis this is the only bending moment considered to produce the elastic curve. In second-order analysis the bending moment due to the product of the axial force and deflection is combined with that due to the transverse loads in producing the elastic curve. Here again, the total bending moment and the elastic curve are mutually implicit.

It suffices to say at this point that only first-order analysis will be considered in this introductory text. To repeat, in first-order analysis, the equilibrium equations are based on the geometry of the original, undeformed shape.

1.3 Linear vs. Nonlinear Material Properties

In considering the response of structural frames to loads or disturbances of any kind, the material properties are certainly the required data. Of primary importance is the stress–strain curve obtained in the laboratory; for instance, of a steel rod in tension or of a concrete cylinder in compression. While the stress–strain curve of steel is, for all practical purposes, linear within a substantial range, that of concrete is curved over its entire length. In addition, when concrete is reinforced with steel and subjected to the development of cracks in the tension zone, the behavior of an entire member can only be assumed statistically. In Fig. 1.3c an idealized stress–strain curve ABC is shown, which can reasonably be used as a basis in the derivation of mathematical methods.

The subjects in this introductory text are limited to dealing with that portion of the structural behavior as governed by the linear portion AB of the idealized stress–strain curve in Fig. 1.3c.

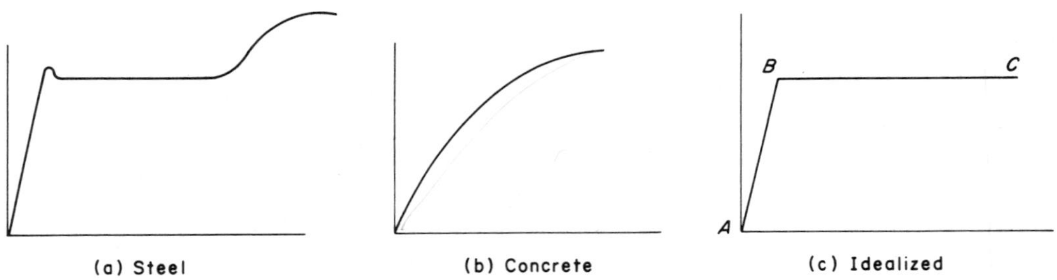

FIG. 1.3. Typical Stress–Strain Curves.

1.4 Static vs. Dynamic Loads

When a force P is gradually applied to a linear elastic spring as shown in Fig. 1.4a, the extension X is at all times proportional to the applied force P such that

$$P = KX \tag{1.4.1}$$

where K is the stiffness of the spring. The work done in effecting an additional extension dX is

$$dW = PdX = KXdX$$

and the total work done in extending the spring length from L_0 to $(L_0 - X)$ is

$$W = \int dW = \int_0^X KXdX = \tfrac{1}{2}KX^2 = \tfrac{1}{2}PX \tag{1.4.2}$$

Thus, the elastic energy stored inside the spring is equal to one half of the product of the resulting force in the spring and its extension.

The behavior of a solid bar, so long as its material properties stay within the linear portion AB of the stress–strain curve of Fig. 1.3c, is no different from that of the linear elastic spring, only that the stiffness K is now

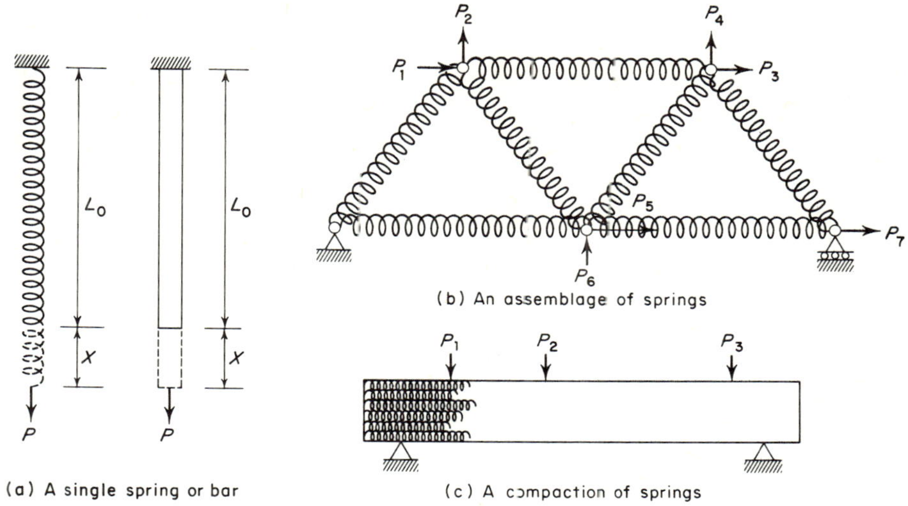

FIG. 1.4. Static Load on Linear Elastic Springs.

$$K = \frac{EA}{L_0} \tag{1.4.3}$$

where E is the modulus of elasticity and A is the cross-sectional area.

If, instead of applying a gradual force to the spring, a mass with weight equal to W is hung at the end, there arises the question of whether the mass is gradually or suddenly released to act on the spring. In the former case, the load is said to be static and the resulting extension, by use of Eq. (1.4.1), is

$$X(\text{static}) = \frac{W}{K} \tag{1.4.4}$$

In the latter case, the load is said to be dynamic and the result of the impact on the spring is an extension twice as great as the static deflection of Eq. (1.4.4) after which there would be oscillatory motion of the mass.

The truss of Fig. 1.4b is considered to consist of seven linear elastic springs hooked together at the five joints. The load system can be completely defined by the seven values of P. If the loads are gradually applied, the seven unknown deflections will also gradually increase from zero to their full amounts; and so likewise, the forces and deformations in the springs. Even the beam of Fig. 1.4c may be considered to consist of an infinite number of little springs compacted together, and its behavior under static loads is similar to that of the single bar or of a group of bars.

In this introductory text, only static loads—gradually applied loads—will be considered. The subject of structural analysis for dynamic loads, particularly in cases where the springs themselves possess mass, has been popularly named *structural dynamics.* In that context, this book is limited to *structural statics.*

1.5 Elastic Instability vs. Statical Instability

When a bar hinged at one end and supported on rollers at the other, as the one shown in Fig. 1.5, is subjected to a compressive axial force P, it should stay straight and the roller support would be displaced downward by an amount equal to

$$X = \frac{P}{K} = \frac{PL_0}{AE} \tag{1.5.1}$$

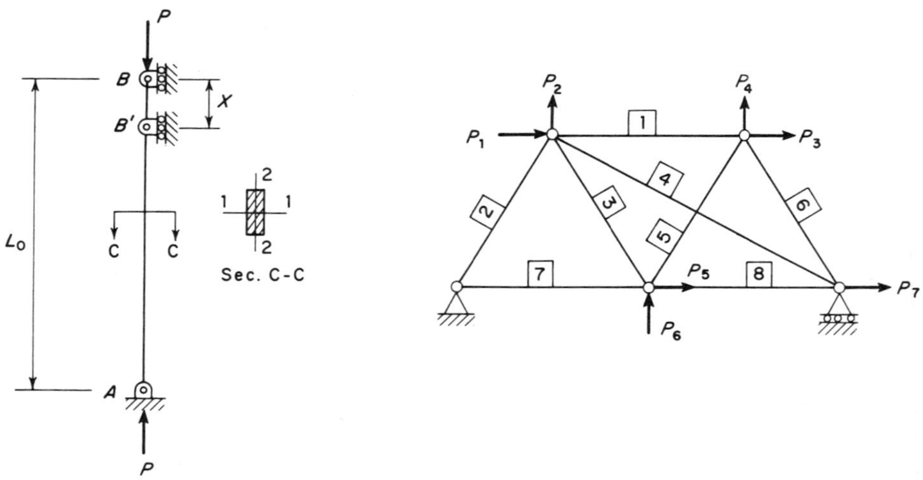

FIG. 1.5. A One-Bar Structure and a Multiple-Bar Truss.

The structural analysis is correct so long as the material property is linear; and, more to the subject of this article, so long as P is below the Euler's buckling load

$$P_{cr} = \frac{\pi^2 EI}{L_0^2} \tag{1.5.2}$$

where I is the moment of inertia about axis 2–2 of the cross section. As derived in textbooks on "Strength of Materials," the Euler's load is one at which the bending moment P_{cr} times y would just cause the transverse deflection y. When the load P is greater than the Euler's load, the one-bar structure of Fig. 1.5 buckles and is termed *elastically instable.*

The structural analysis of the truss in Fig. 1.5b involves the determination of the deflections of all joints and the determination of the axial forces in all the bars. In the event that the axial compressive force in any one bar is larger than its Euler's value indicated by Eq. (1.5.2), the whole set of results would have no physical meaning because this particular bar would have buckled and fallen out of function. In this introductory textbook it is assumed that such elastic instability would not occur, leaving to more advanced treatises the analysis of structures some of whose members become incapacitated due to buckling. However, elastic instability should not be confused with statical instability. For instance, if bar No. 2 in the truss of Fig. 1.5b were omitted, the structure would become incapable of keeping a general load system—an arbitrary set of seven P-values—in statical equilibrium and the structure is termed *statically instable*. Statically instable structures cannot exist and there can be no structural analysis of them.

1.6 Statically Determinate and Statically Indeterminate Structures

As has been previously stated, the structural analysis of a truss like the one in Fig. 1.6a involves the determination of the deflections of all joints and the axial forces in all bars owing to the application of an arbitrary set of P-values. When the entire truss is broken apart into five free bodies of the joints, as shown in Fig. 1.6b, there are available two equations of statical equilibrium per joint or a total of ten conditions. In this particular truss the total number of unknown forces also happens to be ten: three reactions and seven bar forces. When the number of conditions of statics is equal to the number of unknown forces, the structure is said to be *statically determinate* provided it is statically stable. For instance, the truss of Fig. 1.6c may seem to have ten statical conditions and ten unknown forces, yet it would still be statically instable.

The truss of Fig. 1.6d has one more bar than that of Fig. 1.6a; therefore, it still has only ten statical conditions, but has eleven unknown forces. When the number of available independent conditions of statics is less than that of

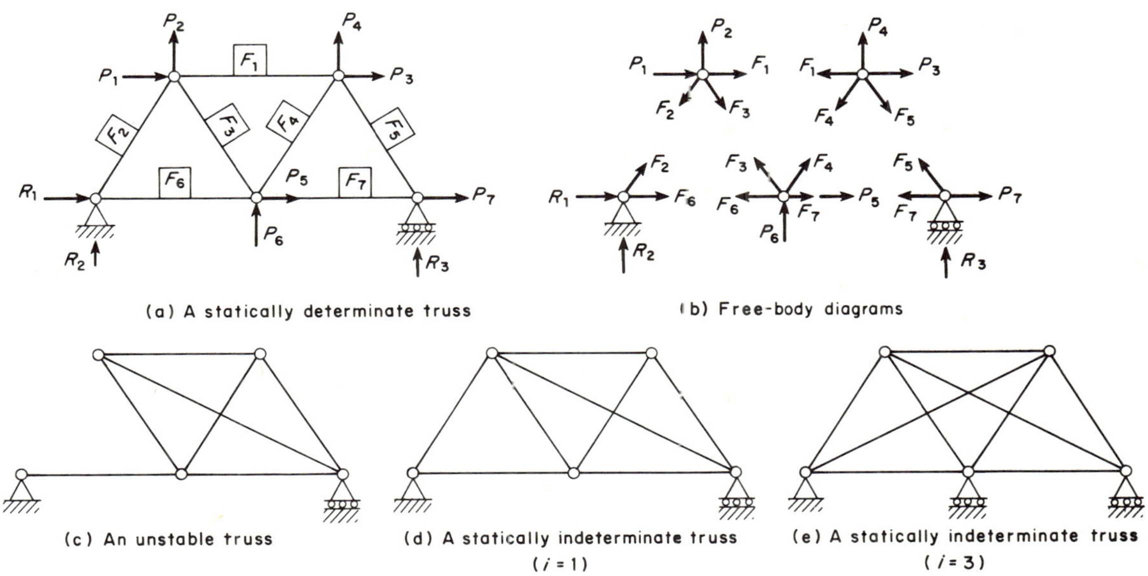

FIG. 1.6. Statically Determinate and Indeterminate Trusses.

unknown forces, the structure is said to be *statically indeterminate*. The number of insufficient conditions of statics is called the *degree of statical indeterminacy*. The degree of statical indeterminacy of the truss in Fig. 1.6d is one, but that of the truss in Fig. 1.6e is three.

Simple, cantilever, and overhanding beams as shown in Fig. 1.7 are statically determinate, but continuous beams in Fig. 1.8 are statically indeterminate to the first, second, and third degrees, respectively. Also, the rigid frames shown in Fig. 1.9 are statically determinate, but those shown in Fig. 1.10 are statically indeterminate. The means of judging the degree of statical indeterminacy of rigid frames will be explained in later chapters.

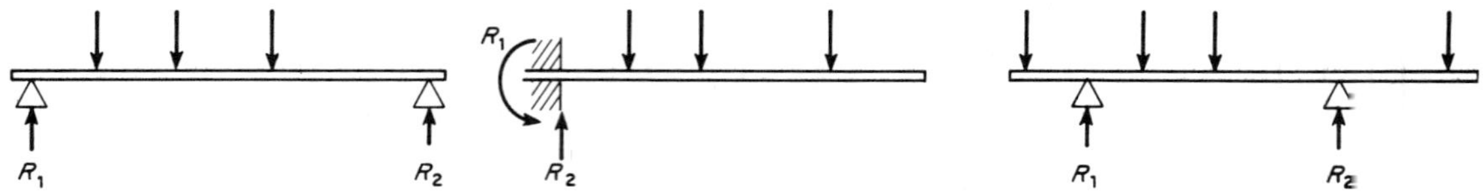

FIG. 1.7. Statically Determinate Beams.

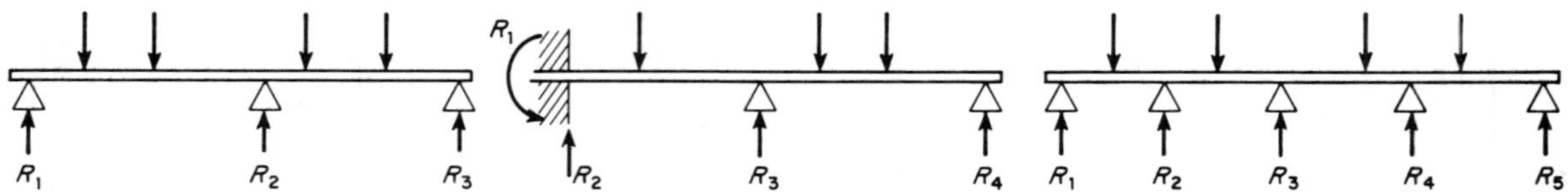

FIG. 1.8. Statically Indeterminate Beams.

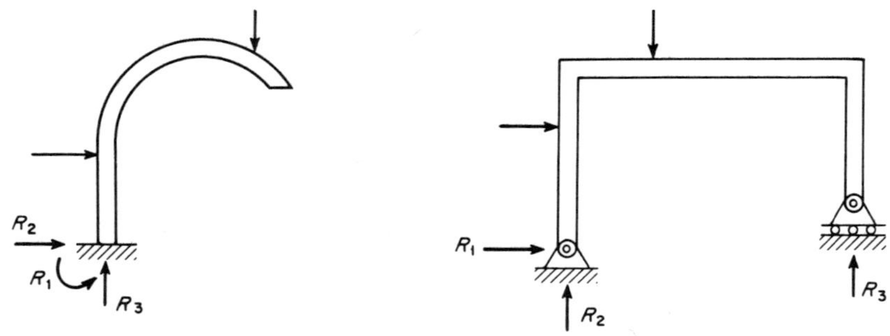

FIG. 1.9. Statically Determinate Rigid Frames.

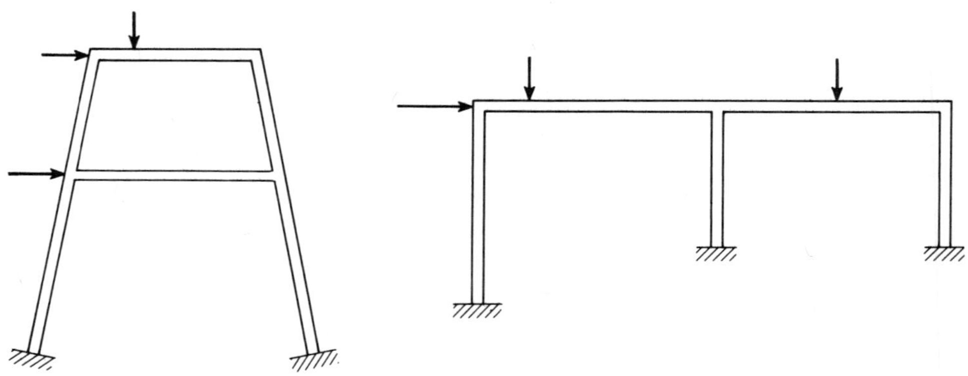

FIG. 1.10. Statically Indeterminate Rigid Frames.

1.7 Displacement Method of Structural Analysis

Consider again the statically determinate truss of Fig. 1.11a and the indeterminate truss of second degree in Fig. 1.11c. Complete knowledge of the behavior of either truss, under an arbitrary set of seven P-values, should include the values of the seven joint displacements in the directions of the P's and the values of the axial forces in the bars, seven in the truss of Fig. 1.11a and nine in that of Fig. 1.11c. If, in setting up the solution, the seven joint displacements are taken as the primary unknowns, it follows:

1) that the change in length of each bar may be expressed by the changes in position of its ends;
2) that the axial force in each bar is expressible by its length change, provided the material property is linear elastic; and
3) that seven equations of joint equilibrium may be written in the seven P-directions using the free bodies as shown in Fig. 1.11b.

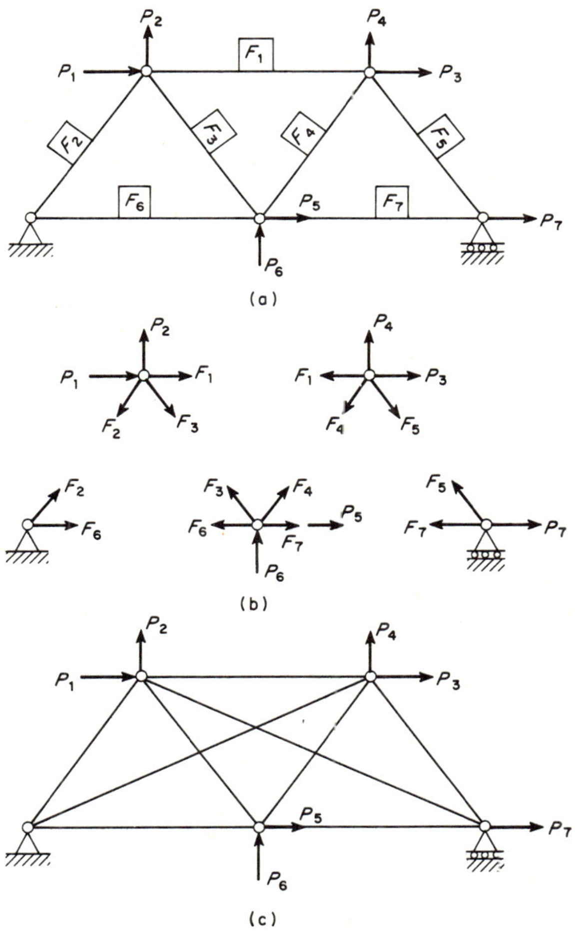

FIG. 1.11. Displacement Method of Truss Analysis.

According to this procedure, regardless of whether the structure is statically determinate or indeterminate, there are always as many equations of statics as unknown joint displacements because there is one load-coordinate for each displacement-coordinate. This method of structural analysis, in which the joint displacements are the primary unknowns, is called the *displacement method.*

The concept of the displacement method is simple, but its application involves the processing of a number of linear transformations and the solution of simultaneous equations. With the help of matrix algebra and electronic computation, these operations now become conveniently feasible. Although a truss analysis problem has been selected to convey the fundamental principle of the displacement method, the method itself is applicable to the analysis of all types of statically determinate and indeterminate structures.

1.8 Force Method of Structural Analysis

The number of unknown forces in the statically determinate truss of Fig. 1.11a is ten, of which three are reactions and seven are axial forces in the bars; but the number of unknown forces in the statically indeterminate truss to the second degree of Fig. 1.11c is twelve, whereas the number of independent equations of statics remains at ten. The ten unknown forces in the statically determinate truss can be solved directly from the ten statical equations, even without knowledge of the sizes and material properties of the seven bars. Thus, although the deformation response of a statically determinate structure must depend on the elastic properties of its constituent members, the internal-force response is dictated entirely by the original shape of the structure.

The situation would be different in the case of the statically indeterminate truss of Fig. 1.11c. If it were insisted that the twelve unknown forces were to be taken as the primary unknowns, two additional conditions to the ten statical conditions would be required. These conditions are those of compatibility of deformations. The concept involves:

1) the cutting of any two bars so long as a statically determinate and statically stable truss remains;
2) the determination of the amount of gap or overlap at the two cuts owing to the applied loads; and
3) the determination of the axial forces in the two cut bars which, when applied in pairs at the cuts, would eliminate the gaps or overlaps.

Thus, the unknown axial forces in the bars selected to be cut may be regarded as *super* primary unknowns, and they must be determined first from the compatibility conditions. This method of structural analysis, in which the redundant forces are determined ahead of all other information, is called the *force method* of structural analysis. In using the force method to analyze statically inderminate structures, the elastic properties of the constituent members are required in the evaluation and then elimination of the gaps or overlaps at the cuts of the derived statically determinate structure.

The force method applies not only to trusses but also to all other types of framed structures. In this introductory text both the displacement and force methods will be forcibly treated.

1.9 Concluding Remarks

An attempt has been made in this first chapter to enumerate the limitations on the subject matter treated in this textbook and then to describe in general the displacement and force methods of structural analysis. The limitations are: 1) framed structures over plate or shell structures; 2) first-order analysis over second-order analysis; 3) linear elastic over nonlinear or inelastic material properties; and 4) static over dynamic loads. In addition, possibility of elastic instability is excluded.

The displacement method, being applicable to both statically determinate and indeterminate structures, is becoming increasingly popular because of its easy adaptation to the electronic computer through matrix formulation. The force method, more aptly named the redundant force method, should rightfully be linked to the analysis of statically indeterminate structures. It is a procedure of considerable scholarly elegance, and traditionally has been the only method of analyzing statically indeterminate trusses. That students of structural theory should understand both of these methods is very desirable.

Matrix Operations

2

2.1 Matrix Form of a System of Linear Equations

Suppose that the relationships between three values of x and four values of y may be expressed by three linear equations as follows:

$$\begin{aligned} x_1 &= 2y_1 + 3y_2 + 4y_3 + 5y_4 \\ x_2 &= 6y_1 + 7y_2 + 8y_3 + 9y_4 \\ x_3 &= 10y_1 + 11y_2 + 12y_3 + 13y_4 \end{aligned} \tag{2.1.1}$$

The 12 constants which appear in Eq. (2.1.1) may be arranged in the form of a *matrix*, which is a rectangular block of numbers; thus

(2.1.2a)
$$[A]_{3\times 4} = \begin{bmatrix} +2 & +3 & +4 & +5 \\ +6 & +7 & +8 & +9 \\ +10 & +11 & +12 & +13 \end{bmatrix}$$

A *rectangular matrix* is denoted by enclosing a capital letter inside square brackets, where the first and second subscripts indicate the number of rows and columns, respectively. Similarly the x or y values may be arranged in the form of a *column matrix*, which is the special name of a rectangular matrix with only one column; thus

(2.1.2b)
$$\{X\}_{3\times 1} = \begin{bmatrix} x_1 \\ x_2 \\ x_3 \end{bmatrix} \qquad \{Y\}_{4\times 1} = \begin{bmatrix} y_1 \\ y_2 \\ y_3 \\ y_4 \end{bmatrix}$$

A column matrix is denoted by enclosing a capital letter inside curved brackets.

The matrix form of a system of linear equations like Eq. (2.1.1) is simply

$$\{X\}_{3\times 1} = [A]_{3\times 4}\{Y\}_{4\times 1}$$

where $\{X\}$, $[A]$, and $\{Y\}$ are as defined in Eq. (2.1.2ab). In general terms, the matrix form of a system of linear equations

(2.1.3)
$$\begin{aligned} x_1 &= a_{11}y_1 + a_{12}y_2 + \cdots a_{1j}y_j + \cdots + a_{1M}y_M \\ x_2 &= a_{21}y_1 + a_{22}y_2 + \cdots a_{2j}y_j + \cdots + a_{2M}y_M \\ &\cdots \\ x_i &= a_{i1}y_1 + a_{i2}y_2 + \cdots a_{ij}y_j + \cdots + a_{iM}y_M \\ &\cdots \\ x_L &= a_{L1}y_1 + a_{L2}y_2 + \cdots a_{Lj}y_j + \cdots + a_{LM}y_M \end{aligned}$$

is

(2.1.4a)
$$\{X\}_{L\times 1} = [A]_{L\times M}\{Y\}_{M\times 1}$$

where

(2.1.4b)
$$\{X\}_{L\times 1} = \begin{bmatrix} x_1 \\ x_2 \\ \cdots \\ x_L \end{bmatrix} \qquad [A]_{L\times M} = \begin{bmatrix} a_{11} & a_{12} & \cdots & a_{1M} \\ a_{21} & a_{22} & \cdots & a_{2M} \\ \cdots & & & \\ a_{L1} & a_{L2} & \cdots & a_{LM} \end{bmatrix} \qquad [Y]_{M\times 1} = \begin{bmatrix} y_1 \\ y_2 \\ \cdots \\ y_M \end{bmatrix}$$

It may be noted that a_{ij} is the element at the ith row and jth column of the $[A]$ matrix.

2.2 Matrix Multiplication

If two systems of linear equations

$$
\begin{aligned}
x_1 &= 2y_1 + 3y_2 + 4y_3 + 5y_4 \\
x_2 &= 6y_1 + 7y_2 + 8y_3 + 9y_4 \\
x_3 &= 10y_1 + 11y_2 + 12y_3 + 13y_4
\end{aligned}
\tag{2.2.1}
$$

and

$$
\begin{aligned}
y_1 &= 14z_1 + 15z_2 \\
y_2 &= 16z_1 + 17z_2 \\
y_3 &= 18z_1 + 19z_2 \\
y_4 &= 20z_1 + 21z_2
\end{aligned}
\tag{2.2.2}
$$

are given, it is possible to derive a third system of linear equations in which the three values of x are expressed directly by the two values of z by substituting Eq. (2.2.2) in Eq. (2.2.1); thus

(2.2.3)

$$
\begin{aligned}
x_1 &= 2(14z_1 + 15z_2) + 3(16z_1 + 17z_2) + 4(18z_1 + 19z_2) + 5(20z_1 + 21z_2) \\
&= 248z_1 + 262z_2 \\
x_2 &= 6(14z_1 + 15z_2) + 7(16z_1 + 17z_2) + 8(18z_1 + 19z_2) + 9(20z_1 + 21z_2) \\
&= 520z_1 + 550z_2 \\
x_3 &= 10(14z_1 + 15z_2) + 11(16z_1 + 17z_2) + 12(18z_1 + 19z_2) + 13(20z_1 + 21z_2) \\
&= 792z_1 + 838z_2
\end{aligned}
$$

Equations (2.2.1) to (2.2.3) may be written in matrix form as

$$\{X\}_{3\times 1} = [A]_{3\times 4}\{Y\}_{4\times 1} \tag{2.2.4}$$

$$\{Y\}_{4\times 1} = [B]_{4\times 2}\{Z\}_{2\times 1} \tag{2.2.5}$$

$$\{X\}_{3\times 1} = [C]_{3\times 2}\{Z\}_{2\times 1} \tag{2.2.6}$$

If Eq. (2.2.5) is substituted in Eq. (2.2.4), the following is obtained:

$$\{X\} = [A]\{Y\} = [A][B]\{Z\} \tag{2.2.7}$$

Comparing Eq. (2.2.7) with Eq. (2.2.6), one may deduce

$$[C]_{3\times 2} = [A]_{4\times 3}[B]_{3\times 2} \tag{2.2.8}$$

The $[C]$ matrix is thus defined as the product of the matrices $[A]$ and $[B]$. It is important to note that the number of columns in the premultiplier $[A]$ must be equal to the number of rows in the postmultiplier $[B]$.

The arithmetic involved in the substitution of Eq. (2.2.2) in Eq. (2.2.1) may be tabulated as shown below.

$[B]_{4\times 2} =$

14	15
16	17
18	19
20	21

$[A]_{3\times 4} =$

2	3	4	5
6	7	8	9
10	11	12	13

$[C]_{3\times 2} =$

2 × 14 = 28 3 × 16 = 48 4 × 18 = 72 5 × 20 = 100 248	2 × 15 = 30 3 × 17 = 51 4 × 19 = 76 5 × 21 = 105 262
6 × 14 = 84 7 × 16 = 112 8 × 18 = 144 9 × 20 = 180 520	6 × 15 = 90 7 × 17 = 119 8 × 19 = 152 9 × 21 = 189 550
10 × 14 = 140 11 × 16 = 176 12 × 18 = 216 13 × 20 = 260 792	10 × 15 = 150 11 × 17 = 187 12 × 19 = 228 13 × 21 = 273 838

where c_{ij} is shown to be the sum of the products of a_{ik} and b_{kj} as k increases from 1 to the number of columns in the premultiplier or the number of rows in the postmultiplier. Thus, the rule for multiplying $[A]_{L\times M}$ and $[B]_{M\times N}$ is

$$C(i, j) = \sum_{k=1}^{k=M} [A(i, k) * B(k, j)] \tag{2.2.9}$$

EXERCISE 2.1

Given

$$[A] = \begin{bmatrix} -1 & +2 & -3 \\ +4 & -5 & -6 \\ +7 & -8 & +9 \end{bmatrix} \quad \text{and} \quad B = \begin{bmatrix} +10 & -11 \\ -12 & +13 \\ -14 & +15 \end{bmatrix},$$

compute $[C] = [A][B]$. Compute $\{X\}$ from $\{X\} = [C]\{Z\}$ directly if $z_1 = +3$ and $z_2 = -2$; and then compute $\{X\}$ from $\{X\} = [A]\{Y\}$ after obtaining $\{Y\}$ from $\{Y\} = [B]\{Z\}$ when $z_1 = +3$ and $z_2 = -2$.

EXERCISE 2.2

Given

$$[A] = \begin{bmatrix} a & 0 & 0 \\ 0 & b & 0 \\ 0 & 0 & c \end{bmatrix} \quad \text{and} \quad [B] = \begin{bmatrix} g & h \\ e & f \\ i & j \end{bmatrix},$$

obtain $[AB]$.

EXERCISE 2.3

Given

$$[A] = \begin{bmatrix} e & f & g \\ h & i & j \end{bmatrix} \quad \text{and} \quad [B] = \begin{bmatrix} a & 0 & 0 \\ 0 & b & 0 \\ 0 & 0 & c \end{bmatrix},$$

obtain $[AB]$.

2.3 Matrix Inversion

Suppose that three values of x are expressible in terms of three values of y by three linear equations as follows:

$$\begin{aligned} x_1 &= 2y_1 + 3y_2 + 4y_3 \\ x_2 &= 5y_1 + 6y_2 + 7y_3 \\ x_3 &= 8y_1 + 9y_2 + 6y_3 \end{aligned} \tag{2.3.1}$$

The matrix form of the system of linear equations (2.3.1) is

(2.3.2) $$\{X\}_{3\times 1} = [A]_{3\times 3}\{Y\}_{3\times 1}$$

in which

(2.3.3) $$[A]_{3\times 3} = \begin{bmatrix} +2 & +3 & +4 \\ +5 & +6 & +7 \\ +8 & +9 & +6 \end{bmatrix}$$

Unless there is dependency in the three equations of (2.3.1), it is pcssible to express the three values of y in terms of the three values of x by

(2.3.4) $$\begin{aligned} y_1 &= b_{11}x_1 + b_{12}x_2 + b_{13}x_3 \\ y_2 &= b_{21}x_1 + b_{22}x_2 + b_{23}x_3 \\ y_3 &= b_{31}x_1 + b_{32}x_2 + b_{33}x_3 \end{aligned}$$

The matrix form of the system of linear equations (2.3.4) is

(2.2.5) $$\{Y\}_{3\times 1} = [B]_{3\times 3}\{X\}_{3\times 1}$$

in which

(2.3.6) $$[B]_{3\times 3} = \begin{bmatrix} b_{11} & b_{12} & b_{13} \\ b_{21} & b_{22} & b_{23} \\ b_{31} & b_{32} & b_{33} \end{bmatrix}$$

The matrix $[B]$ is defined as the inverse of the matrix $[A]$; in symbolic form,

(2.3.7) $$[B] = [A]^{-1} = [A^{-1}]$$

Equation (2.3.5) may now be written as

(2.3.8) $$\{Y\} = [A^{-1}]\{X\}$$

Substituting Eq. (2.3.8) in Eq. (2.3.2)

(2.3.9) $$\{X\} = [A]\{Y\} = [A][A^{-1}]\{X\} = [AA^{-1}]\{X\}$$

In conventional form, Eq. (2.3.9) must be

(2.3.10) $$\begin{aligned} x_1 &= 1.0\,x_1 + 0\quad x_2 + 0\quad x_3 \\ x_2 &= 0\quad x_1 + 1.0\,x_2 + 0\quad x_3 \\ x_3 &= 0\quad x_1 + 0\quad x_2 + 1.0\,x_3 \end{aligned}$$

which means that

$$[AA^{-1}]_{3\times 3} = \begin{bmatrix} 1 & 0 & 0 \\ 0 & 1 & 0 \\ 0 & 0 & 1 \end{bmatrix} = [I]_{3\times 3} \tag{2.3.11}$$

The symbol $[I]$ is to represent a *unit matrix* or an *identity matrix*, which is a square matrix having the value of 1 on the main diagonal and the value of 0 elsewhere. Thus the product of a matrix and its own inverse, which exists only if there is no dependency in the original matrix, is a unit matrix. A matrix containing dependency is called a *singular matrix*.

One of the ways to find the inverse of a matrix is to solve the simultaneous equations obtained by interchanging the left and right sides of Eq. (2.3.1); thus

$$\begin{aligned} 2y_1 + 3y_2 + 4y_3 &= x_1 \\ 5y_1 + 6y_2 + 7y_3 &= x_2 \\ 8y_1 + 9y_2 + 6y_3 &= x_3 \end{aligned} \tag{2.3.12}$$

Solving the three equations in (2.3.12),

$$\begin{aligned} y_1 &= -(\tfrac{9}{4})x_1 + (\tfrac{3}{2})x_2 - (\tfrac{1}{4})x_3 \\ y_2 &= +(\tfrac{13}{6})x_1 - (\tfrac{5}{3})x_2 + (\tfrac{1}{2})x_3 \\ y_3 &= -\ (\tfrac{1}{4})x_1 + (\tfrac{1}{2})x_2 - (\tfrac{1}{4})x_3 \end{aligned} \tag{2.3.13}$$

This solution can best be found by letting successively ($x_1 = +1$, $x_2 = 0$, $x_3 = 0$), ($x_1 = 0$, $x_2 = +1$, $x_3 = 0$), and ($x_1 = 0$, $x_2 = 0$, $x_3 = +1$) in order that the coefficients in the 1st, 2nd, and 3rd columns of Eq. (2.3.13) may be obtained separately. Thus the inverse of the matrix $[A]$ in Eq. (2.3.3) is

$$[A]^{-1} = \begin{bmatrix} +2 & +3 & +4 \\ +5 & +6 & +7 \\ +8 & +9 & +6 \end{bmatrix}^{-1} = \begin{bmatrix} -\frac{9}{4} & +\frac{3}{2} & -\frac{1}{4} \\ +\frac{13}{6} & -\frac{5}{3} & +\frac{1}{2} \\ -\frac{1}{4} & +\frac{1}{2} & -\frac{1}{4} \end{bmatrix}$$

A check should be made by verifying that

$$[AA^{-1}] = \begin{bmatrix} +2 & +3 & +4 \\ +5 & +6 & +7 \\ +8 & +9 & +6 \end{bmatrix} \begin{bmatrix} -\frac{9}{4} & +\frac{3}{2} & -\frac{1}{4} \\ +\frac{13}{6} & -\frac{5}{3} & +\frac{1}{2} \\ -\frac{1}{4} & +\frac{1}{2} & -\frac{1}{4} \end{bmatrix} = \begin{bmatrix} 1 & 0 & 0 \\ 0 & 1 & 0 \\ 0 & 0 & 1 \end{bmatrix} = [I] \tag{2.3.14}$$

2.4 Solution of Linear Simultaneous Equations

One of the ways of solving linear simultaneous equations is through the Gauss–Jordan Elimination Method. In this method the solution of a system of equations such as

$$\begin{aligned} a_{11}x_1 + a_{12}x_2 + a_{13}x_3 &= a_{14} \\ a_{21}x_1 + a_{22}x_2 + a_{23}x_3 &= a_{24} \\ a_{31}x_1 + a_{32}x_2 + a_{33}x_3 &= a_{34} \end{aligned} \tag{2.4.1}$$

is effected by transforming the original matrix of coefficients through three eliminations to the final matrix as shown below.

$$\begin{bmatrix} a_{11} & a_{12} & a_{13} & a_{14} \\ a_{21} & a_{22} & a_{23} & a_{24} \\ a_{31} & a_{32} & a_{33} & a_{34} \end{bmatrix} \longrightarrow \begin{bmatrix} 1 & a'_{12} & a'_{13} & a'_{14} \\ 0 & a'_{22} & a'_{23} & a'_{24} \\ 0 & a'_{32} & a'_{33} & a'_{34} \end{bmatrix} \longrightarrow$$

(1st elimination)

$$\longrightarrow \begin{bmatrix} 1 & 0 & a''_{13} & a''_{14} \\ 0 & 1 & a''_{23} & a''_{24} \\ 0 & 0 & a''_{33} & a''_{34} \end{bmatrix} \longrightarrow \begin{bmatrix} 1 & 0 & 0 & a'''_{14} \\ 0 & 1 & 0 & a'''_{24} \\ 0 & 0 & 1 & a'''_{34} \end{bmatrix}$$

(2nd elimination) (3rd elimination)

The final matrix yields the solution

$$\{X\} = \begin{bmatrix} x_1 \\ x_2 \\ x_3 \end{bmatrix} = \begin{bmatrix} a'''_{14} \\ a'''_{24} \\ a'''_{34} \end{bmatrix}$$

The ith row at the end of the ith elimination is called the *pivotal row* whose elements are obtained by dividing the ith row in the previous matrix by a_{ii} (the pivot) in that matrix. The rows other than the ith row at the end of the ith elimination are obtained by subtracting, from the corresponding row in the previous matrix, the product of a constant and the pivotal row (the ith row) with the purpose of attaining the value of zero directly above or below the value of 1 in the pivotal row.

As an illustration, the columns of the inverse of

$$[A] = \begin{bmatrix} +2 & +3 & +4 \\ +5 & +6 & +7 \\ +8 & +9 & +6 \end{bmatrix}$$

in the preceding article may be obtained by solving three systems of three linear equations

$$
\begin{aligned}
2y_1 + 3y_2 + 4y_3 &= 1, \quad 0, \quad 0 \\
5y_1 + 6y_2 + 7y_3 &= 0, \quad 1, \quad 0 \\
8y_1 + 9y_2 + 6y_3 &= 0, \quad 0, \quad 1
\end{aligned}
$$

The Gauss–Jordan eliminations are, in succession,

$$
\begin{bmatrix} +2 & +3 & +4 & +1 & 0 & 0 \\ +5 & +6 & +7 & 0 & +1 & 0 \\ +8 & +9 & +6 & 0 & 0 & +1 \end{bmatrix} \longrightarrow \begin{bmatrix} +1 & +\frac{3}{2} & +2 & +\frac{1}{2} & 0 & 0 \\ 0 & -\frac{3}{2} & -3 & -\frac{5}{2} & +1 & 0 \\ 0 & -3 & -10 & -4 & 0 & +1 \end{bmatrix} \longrightarrow
$$

(1st elimination)

$$
\longrightarrow \begin{bmatrix} +1 & 0 & -1 & -2 & +1 & 0 \\ 0 & +1 & +2 & +\frac{5}{3} & -\frac{2}{3} & 0 \\ 0 & 0 & -4 & +1 & -2 & +1 \end{bmatrix} \longrightarrow \begin{bmatrix} +1 & 0 & 0 & -\frac{9}{4} & +\frac{3}{2} & -\frac{1}{4} \\ 0 & +1 & 0 & +\frac{13}{6} & -\frac{5}{3} & +\frac{1}{2} \\ 0 & 0 & +1 & -\frac{1}{4} & +\frac{1}{2} & -\frac{1}{4} \end{bmatrix}
$$

(2nd elimination) (3rd elimination)

Thus,

$$
\begin{bmatrix} +2 & +3 & +4 \\ +5 & +6 & +7 \\ +8 & +9 & +6 \end{bmatrix}^{-1} = \begin{bmatrix} -\frac{9}{4} & +\frac{3}{2} & -\frac{1}{4} \\ +\frac{13}{6} & -\frac{5}{3} & +\frac{1}{2} \\ -\frac{1}{4} & +\frac{1}{2} & -\frac{1}{4} \end{bmatrix}
$$

Note, for example, that the second row at first elimination $(0, -\frac{3}{2}, -3, -\frac{5}{2}, +1, 0)$ is obtained by subtracting $+5$ times the pivotal row $(+1, +\frac{3}{2}, +2, +\frac{1}{2}, 0, 0)$ from the second row in the original matrix $(+5, +6, +7, 0, +1, 0)$. Likewise, the third row $(0, -3, -10, -4, 0, +1)$ is obtained by subtracting $+8$ times the pivotal row $(+1, +\frac{3}{2}, +2, +\frac{1}{2}, 0, 0)$ from the third row in the original matrix $(+8, +9, +6, 0, 0, +1)$.

In this solution the pivots are taken successively from a'_{11}, through a'_{22}, to a''_{33}. There are many variations in the procedure involving reshuffling the rows to improve the accuracy of the solution. The reader is referred to standard texts on numerical analysis for a more complete exposition.

As discussed in the preceding article, a square matrix with dependency in it has no inverse and is called a singular matrix. The matrix shown below is singular because at the end of second elimination, the third row reads

$$
0y_1 + 0y_2 + 0y_3 = +1, \quad -2, \quad +1
$$

which is impossible.

$$
\begin{bmatrix} +2 & +3 & +4 & +1 & 0 & 0 \\ +5 & +6 & +7 & 0 & +1 & 0 \\ +8 & +9 & +10 & 0 & 0 & +1 \end{bmatrix} \longrightarrow \begin{bmatrix} +1 & +\frac{3}{2} & +2 & +\frac{1}{2} & 0 & 0 \\ 0 & -\frac{3}{2} & -3 & -\frac{5}{2} & +1 & 0 \\ 0 & -3 & -10 & -4 & 0 & +1 \end{bmatrix} \longrightarrow
$$

(1st elimination)

$$
\longrightarrow \begin{bmatrix} +1 & 0 & -1 & -2 & +1 & 0 \\ 0 & +1 & +2 & +\frac{5}{3} & -\frac{2}{3} & 0 \\ 0 & 0 & 0 & +1 & -2 & +1 \end{bmatrix}
$$

(2nd elimination)

Actually in this case the dependency is obvious because the difference between the first row and the second row in the original matrix is equal to that between the second row and the third row. Also, it is shown in matrix algebra that the determinant of a singular matrix is zero; thus

$$
\begin{bmatrix} +2 & +3 & +4 \\ +5 & +6 & +7 \\ +8 & +9 & +10 \end{bmatrix} = (+2)(+6)(+10) + (+5)(+9)(+4) + (+8)(+3)(+7) \\ -(+8)(+6)(+4) - (+5)(+3)(+10) - (+2)(+9)(+7) \\ = 120 + 180 + 168 - 192 - 150 - 126 = 0
$$

EXERCISE 2.4

By the Gauss–Jordan elimination method, obtain the inverse of the matrix

$$
\begin{bmatrix} -\frac{9}{4} & +\frac{3}{2} & -\frac{1}{4} \\ +\frac{13}{6} & -\frac{5}{3} & +\frac{1}{2} \\ -\frac{1}{4} & +\frac{1}{2} & -\frac{1}{4} \end{bmatrix}
$$

EXERCISE 2.5

By the Guass–Jordan elimination method, obtain the inverse of the matrix

$$
\begin{bmatrix} +a & +b \\ +b & +c \end{bmatrix}
$$

2.5 Concluding Remarks

It can be expected that many students would have had a course in matrix or linear algebra before this textbook is studied. However, the barest minimum of matrix definitions, matrix multiplication, and matrix inversion is introduced here in the eventuality that the reader as yet has had no background in matrices. Only the essentials needed for the study of the next several chapters have been included.

The students are urged to work through all five exercises in this chapter. Exercises 2.2 and 2.3 show the effect of a *diagonal matrix* (a square matrix in which the only nonzero elements are on the diagonal) when used as a premultiplier or a postmultiplier. The inverse of the matrix in Exercise 2.4 should be the original matrix as used in the discussion of Art. 2.3. Note or memorize in particular the inverse of the matrix in Exercise 2.5 because this will be useful later in the treatment of the stiffness and flexibility matrices of a member in bending.

Statically Determinate Trusses

3

3.1 Definition of a Statically Determinate Truss

A *truss* is a framed structure consisting of a finite number of bars connected at frictionless pin joints at which external forces may be applied. A typical truss joint in the ideal situation is shown in Fig. 3.1a where a pin is inserted at the center of the eyed ends of the bars. The bars are free to turn at the pin. This is the basis of the analysis treated in this chapter. If the bars are welded together, the joint becomes a *rigid joint*; then the framed structure is a *rigid frame*, not a truss.

So long as the external forces are applied only at the joints, each bar in a truss can be subjected only to a pair of equal and opposite axial forces; thus, the name *two-force members*. This fact can be confirmed by considering the bar proper without the joints, as shown in Fig. 3.1b. Thus the unknown internal

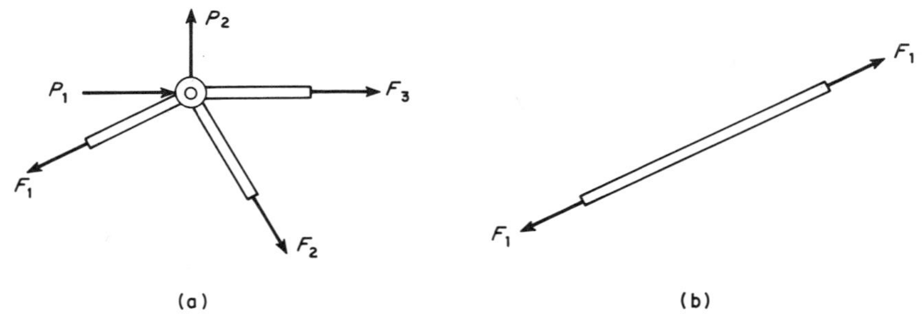

FIG. 3.1. A Typical Joint and Bar in a Truss.

forces are the axial forces F_j, for $j = 1$ to $j = NF$, where NF is the total number of bars.

To orient a two-dimensional truss in its own plane, there must be at least one hinged support and one roller support; or as an unusual alternative, three roller supports having no more than two parallel reactions. At a hinged support, there are two unknown components of the hinge reaction; and for a roller support, there is one unknown reaction. Thus the total number of unknown reaction components NR must be equal to or greater than three.

Any external force system acting on a two-dimensional truss may be described by the numerical magnitudes in the two reference directions at all the joints, excepting those along which there are already unknown reaction components. If NP is the total number of possible directions in which the external forces may act, and NJ is the total number of joints, then

$$NP + NR = 2(NJ) \tag{3.1.1}$$

Since the total number of available equations of static equilibrium for a two-dimensional truss is $2(NJ)$ and the total number of unknown forces is $(NF + NR)$, the necessary requirements for statical determinacy is

$$NF + NR = 2(NJ) \tag{3.1.2}$$

Comparing Eqs. (3.1.1) and (3.1.2), one obtains

$$NP = NF \tag{3.1.3}$$

Equation (3.1.3) states that, for a statically determinate truss, the number of possible directions in which external forces may act is equal to the number of bars.

As illustrations, all the four trusses of Fig. 3.2 satisfy the condition $NP = NF$, although only the first three are statically stable.

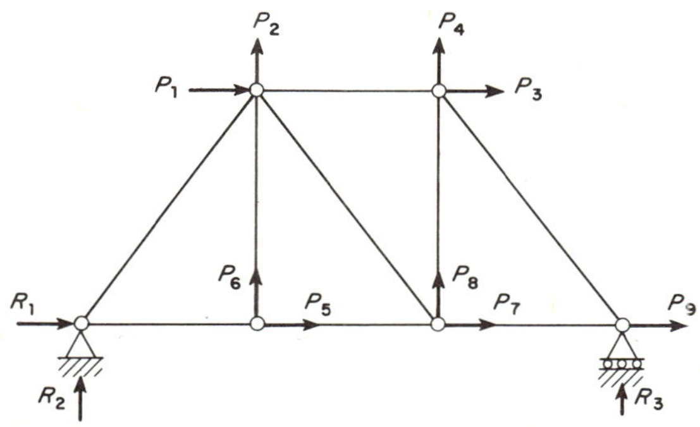

(a) $NP = 9$, $NR = 3$, $NF = 9$; stable

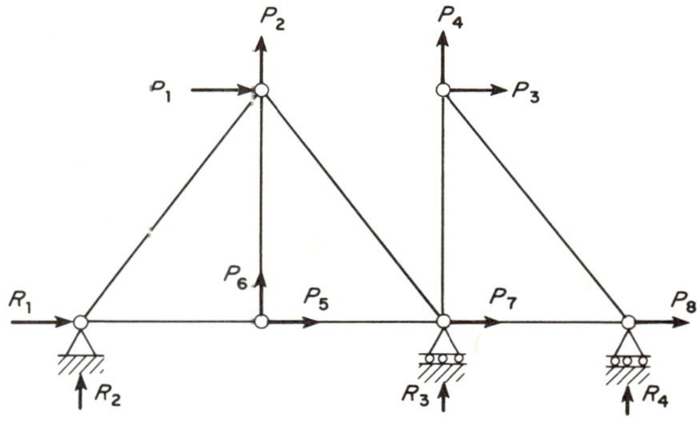

(b) $NP = 8$, $NR = 4$, $NF = 8$; stable

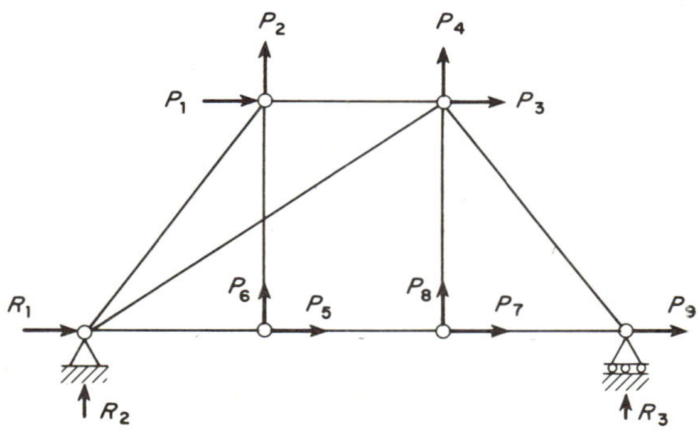

(c) $NP = 9$, $NR = 3$, $NF = 9$; stable

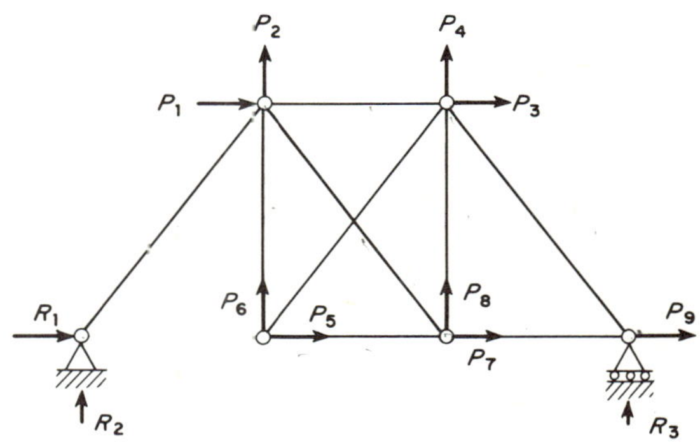

(d) $NP = 9$, $NR = 3$, $NF = 9$; unstable

FIG. 3.2. Typical Statically Determinate Trusses. [The truss in Part (d) is instable.]

3.2 Conventional Method of Joints and Sections

The problem comprising the analysis of a truss is the determination of the external reaction components and the internal axial forces for any given external-force system. Since only statically determinate and stable trusses are treated in this chapter, it is first necessary to ascertain these two requirements. The algebraic condition $NP = NF$ as described in the preceding article is a requirement for statical determinacy but will not detect statical instability. This latter characteristic can usually be recognized by an experienced analyst at the outset while a beginner might not foresee it until he encounters difficulty in completing the analysis.

The conventional (non-computer) method of analyzing statically determinate trusses usually begins with the determination of the reaction components. If the total number is three, this is accomplished by taking the entire truss as a free body. Afterwards, the analyst will search for each successive joint where only two unknown axial forces exist and solve them by using that joint as the free body. The procedure continues until all internal axial forces are computed; thus, the name, *method of joints*. Since the total number of joint equations of equilibrium is three more than the number of unknown axial forces, there will be three surplus equations for checking purposes after all the axial forces have been obtained.

When a visual search does not reveal a subsequent joint where only two unknown bar forces remain, the next avenue open to the analyst is to look for a group of joints together (called a *section of the truss*) upon which there are no more than three unknown bar forces acting, and to solve these by taking this section as the free body; thus, the name, *method of sections*. In many cases the method of sections may be used in lieu of the method of joints, regardless of whether joints with two unknown bar forces are present.

If the total number of reaction components is more than three, and yet the truss is statically determinate, the analyst will have to search for joints with two unknown bar forces or sections (groups of joints) with three unknown bar forces and start from there. In any event a grand final check of the correctness of the solution should be made by placing all axial forces with their hori-

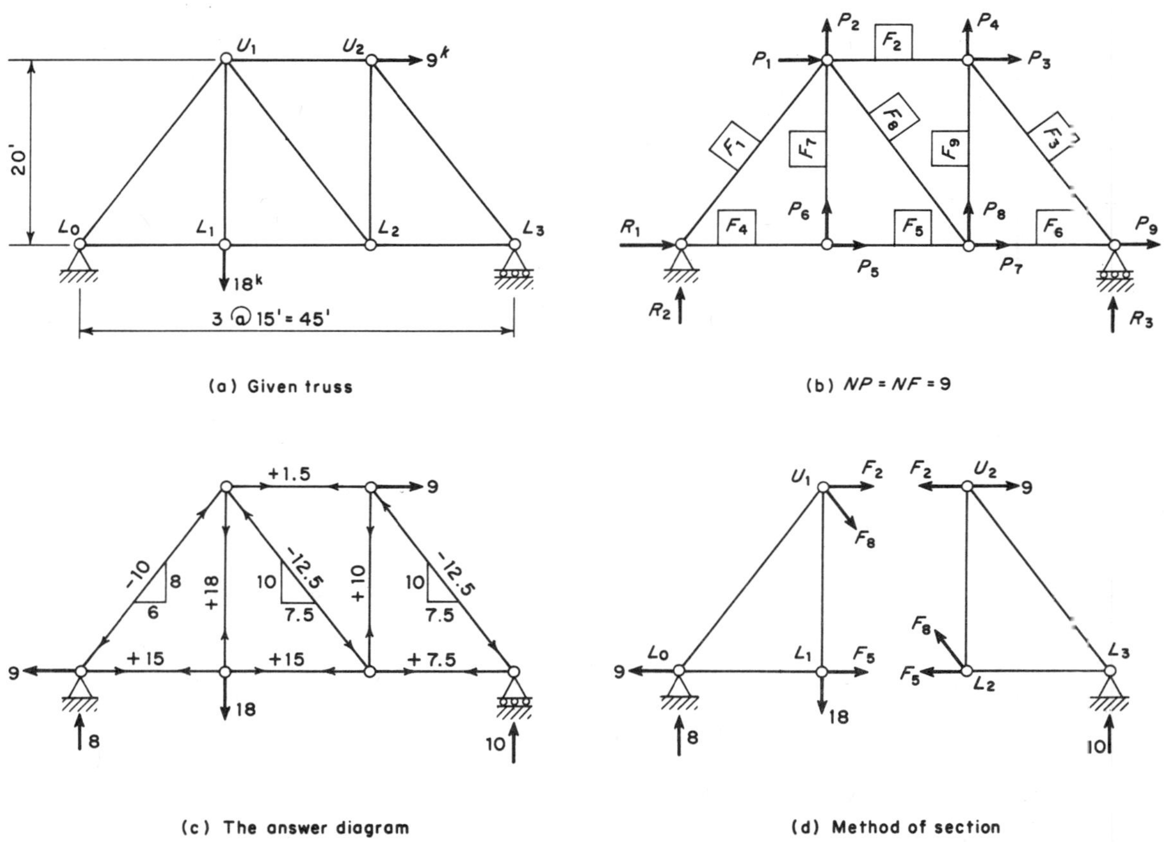

FIG. 3.3. The Truss of Examples 3.1 and 3.2.

zontal and vertical components on a truss diagram and ascertaining that the two equations of statics, $\Sigma F_x = 0$ and $\Sigma F_y = 0$, are satisfied at every joint.

Example 3.1. Analyze the truss of Fig. 3.3a by noting first $NP = NF$; then solving for the reaction components; and lastly, computing the axial forces in all bars by the method of joints.

Solution. From Fig. 3.3b, $NP = NF = 9$. The values shown in the answer diagram of Fig. 3.3c are obtained by the respective steps shown in Table 3.1. Note that Fig. 3.3c is actually the composite free-body diagram of all the joints. Thus, if a bar is pulling on the joint, it is in tension; and if it is pushing, it is in compression.

Example 3.2. Assuming that the reaction components acting on the truss of Fig. 3.3a have already been computed, determine F_2, F_5, and F_8 directly by the method of sections.

Solution. The entire truss may be split into two sections, one containing joints L_0-U_1-L_1 and the other containing joints L_3-L_2-U_2, as shown in Fig. 3.3d. Using the left section as the free body,

1) ΣM about $L_2 = 0$,

$$+8(30) - 18(15) + F_2(20) = 0$$

$$F_2 = +1.5 \text{ kips}$$

TABLE 3.1 Computations for Fig. 3.3c.

Free body	$\Sigma F_x = 0$	$\Sigma F_y = 0$	$\Sigma M = 0$
Entire truss	(2) $R_1 = -9$	(3) $R_2 = +8$	(1) ΣM about L_0, $R_3 = +10$ (4) Check by ΣM about L_3.
Joint U_1	(10) $F_2 = +1.5$	(9) $F_8 = -12.5$	
Joint U_2	(15) Check No. 2	(16) Check No. 3	
Joint L_0	(6) $F_4 = +15$	(5) $F_1 = -10$	
Joint L_1	(8) $F_5 = +15$	(7) $F_7 = +18$	
Joint L_2	(14) Check No. 1	(13) $F_9 = +10$	
Joint L_3	(12) $F_6 = +7.5$	(11) $F_3 = -12.5$	

Note that "point" L_2 is not even on the free body but it is the point of intersection of the lines of action of F_5 and F_8.

2) ΣM about $U_1 = 0$,

$$+9(20) + 8(15) - F_5(20) = 0$$
$$F_5 = +15 \text{ kips}$$

3) $\Sigma F_y = 0$,

$$+8 - 18 - 0.8F_8 = 0$$
$$F_8 = -12.5 \text{ kips}$$

Using the right section as the free body,

4) ΣM about $L_2 = 0$,

$$-10(15) + 9(20) - F_2(20) = 0$$
$$F_2 = +1.5 \text{ kips}$$

5) ΣM about $U_1 = 0$,

$$-10(30) + F_5(20) = 0$$
$$F_3 = +15 \text{ kips}$$

6) $\Sigma F_y = 0$,

$$+10 + 0.8F_8 = 0$$
$$F_8 = -12.5 \text{ kips}$$

The method of sections is a convenient method for directly determining the axial force in any one particular bar in a complex truss if that truss can be split into two sections by cutting through no more than three bars.

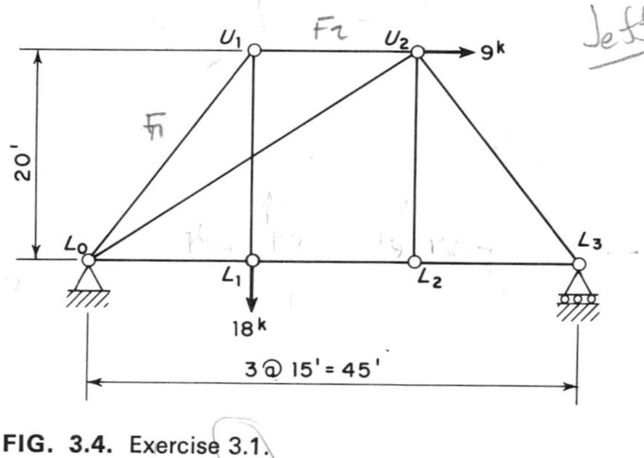

FIG. 3.4. Exercise 3.1.

EXERCISES 3.1 to 3.4

Analyze the statically determinate trusses shown by the method of joints. Choose some bars and determine the axial forces in them by the method of sections (Figs. 3.4–3.7).

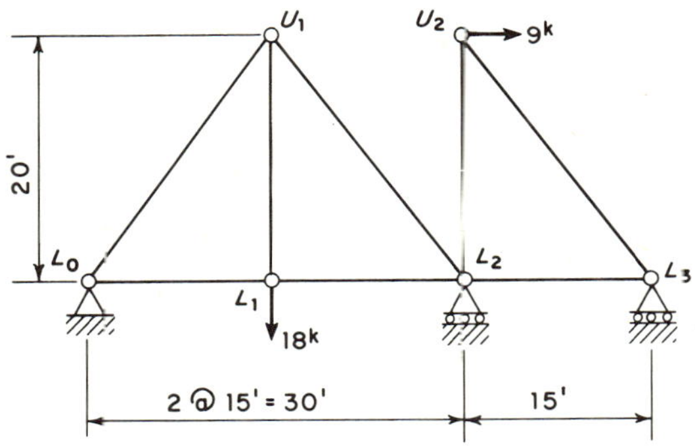

FIG. 3.5. Exercise 3.2.

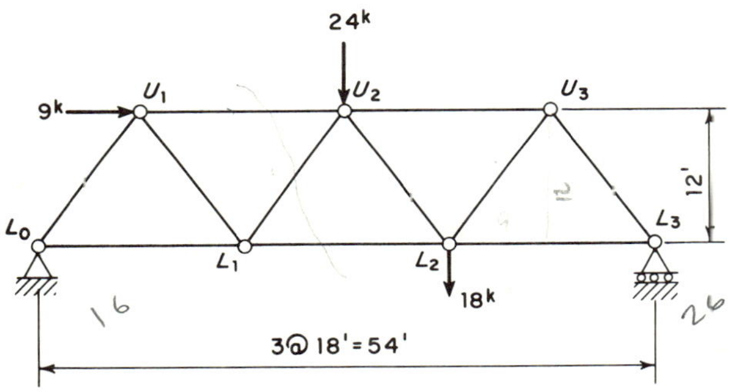

FIG. 3.6. Exercise 3.3.

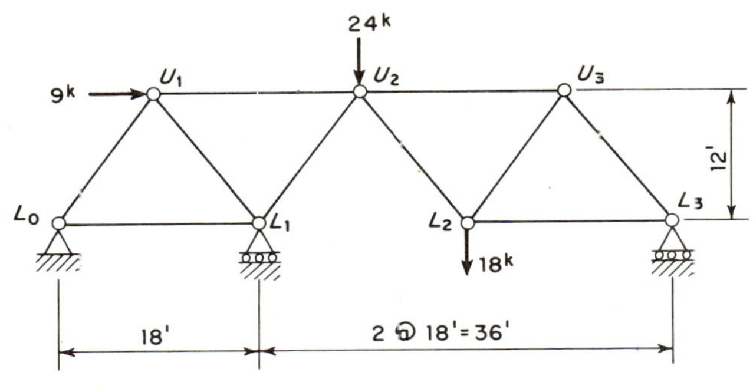

FIG. 3.7. Exercise 3.4.

3.3 Matrix Method of Joints

It has been shown that, for a statically determinate truss, the number of possible directions in which external forces may act, NP, must be equal to the number of axial forces, NF. Let $P(1)$ to $P(NP)$ be the known external joint forces and $F(1)$ to $F(NF)$ be the unknown internal axial forces. Connecting these two sets of variables are NP equations of statics, one each in each direction of P, as follows:

$$P(I) = A(I, 1) * F(1) + A(I, 2) * F(2) + \cdots + A(I, NF) * F(NF) \quad \text{for } I = 1 \text{ to } I = NP \tag{3.3.1}$$

In matrix form, Eq. (3.3.1) becomes

$$\{P\}_{NP\times 1} = [A]_{NP\times NF}\{F\}_{NF\times 1} \tag{3.3.2}$$

The $[A]$ matrix as defined in Eq. (3.3.2) is called the *statics matrix* because its rows are the equations of statics for the joints.

So long as the $[A]$ matrix is square, as it must be for a statically determinate truss, Eq. (3.3.2) may be written as

$$\{F\}_{NF\times 1} = [A^{-1}]_{NF\times NP}\{P\}_{NP\times 1} \tag{3.3.3}$$

This last equation may be programmed on a digital computer with the external-force matrix $\{P\}$ and the statics matrix $[A]$ as input, and the internal-force matrix $\{F\}$ as output.

The physical meaning of the $[A^{-1}]$ matrix is that each of its columns gives

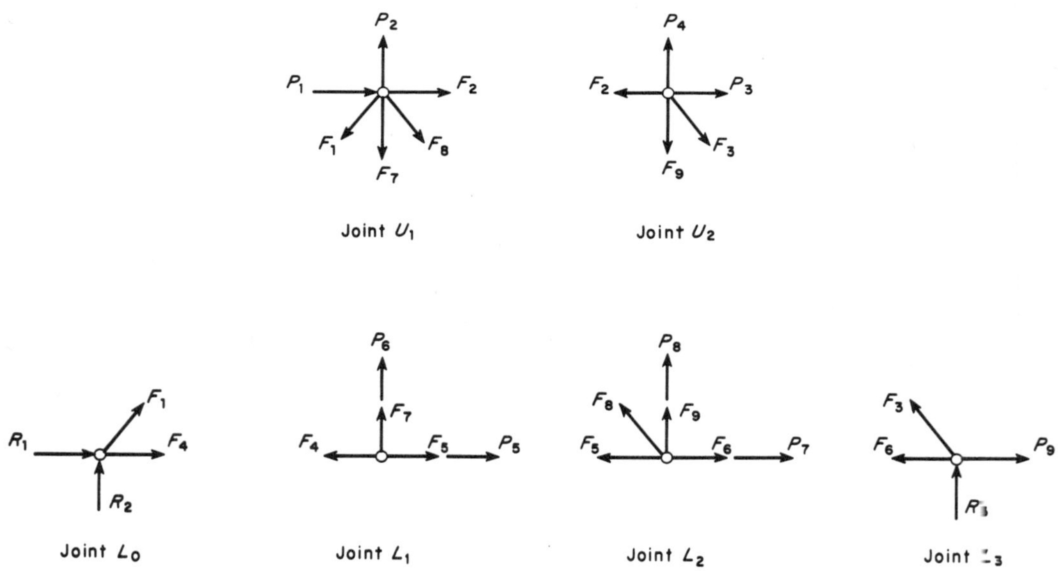

FIG. 3.8. Free-Body Diagrams of Joints in Example 3.3.

the axial forces due to each unit external force. For sake of an academic exercise, the $[A^{-1}]$ matrix may be determined directly by the conventional method of joints and sections, and its correctness verified by confirming that $[A][A^{-1}] = [I]$.

If there exists an inverse of the statics matrix $[A]$, the truss is statically stable, but if the statics matrix $[A]$ is singular, the truss is statically instable.

Example 3.3. For the statically determinate truss of Fig. 3.3a and b; establish the statics matrix $[A]$, determine the $[A^{-1}]$ matrix directly by its physical meaning, and verify the relationship $[A][A^{-1}] = [I]$; then solve for the internal-force matrix $\{F\}$ for the known $\{P\}$ matrix from the equation $\{F\} = [A^{-1}]\{P\}$.

Solution. The statics matrix $[A]$ is simply a compilation of the equations of statics. For instance, using $\Sigma F_x = 0$ and $\Sigma F_y = 0$ for the free-body diagram of Joint U_1 in Fig. 3.8,

$$P_1 = +0.6F_1 - 1.0F_2 - 0.6F_8$$

$$P_2 = +0.8F_1 + 1.0F_7 + 0.8F_8$$

$[A]_{9\times 9} =$

P \ F	1	2	3	4	5	6	7	8	9
1	+0.6	−1.0						−0.6	
2	+0.8						+1.0	+0.8	
3		+1.0	−0.6						
4			+0.8						+1.0
5				+1.0	−1.0				
6							−1.0		
7					+1.0	−1.0		+0.6	
8								−0.8	−1.0
9			+0.6			+1.0			

Each column of the $[A^{-1}]$ matrix simply shows the bar forces due to each unit value of P. For instance, the second and third columns in the $[A^{-1}]$ matrix are taken from the axial force diagrams of Fig. 3.9.

The verification $[A][A^{-1}] = [I]$ was made, although not shown. Also, the internal-force matrix $\{F\}$ computed from $\{F\} = [A^{-1}]\{P\}$ checks exactly with the values on the answer diagram of Fig. 3.3c.

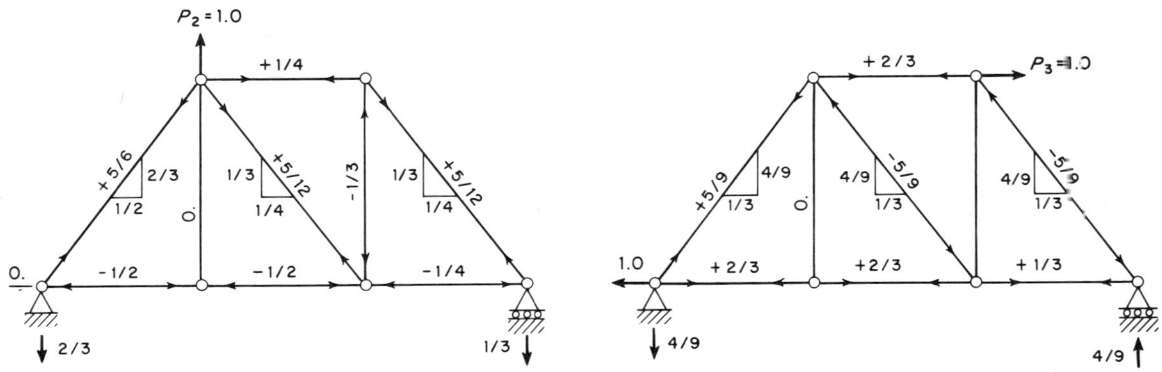

FIG. 3.9. Axial Forces Due to Unit External Forces in Example 3.3.

EXERCISES 3.5 and 3.6

For each of the statically determinate trusses in Exercises 3.1 and 3.2 (Figs. 3.6 and 3.7): establish the statics matrix $[A]$, determine the $[A^{-1}]$ matrix directly by its physical meaning, and verify the relationship $[A][A^{-1}] = [I]$; then solve for the internal-force matrix $\{F\}$ for the known $\{P\}$ matrix from the equation $\{F\} = [A^{-1}]\{P\}$.

$[A^{-1}]_{9\times 9} =$

F \ P	1	2	3	4	5	6	7	8	9
1	$+\frac{5}{9}$	$+\frac{5}{6}$	$+\frac{5}{9}$	$+\frac{5}{12}$	0	$+\frac{5}{6}$	0	$+\frac{5}{12}$	0
2	$-\frac{1}{3}$	$+\frac{1}{4}$	$+\frac{2}{3}$	$+\frac{1}{2}$	0	$+\frac{1}{4}$	0	$+\frac{1}{2}$	0
3	$-\frac{5}{9}$	$+\frac{5}{12}$	$-\frac{5}{9}$	$+\frac{5}{6}$	0	$+\frac{5}{12}$	0	$+\frac{5}{6}$	0
4	$+\frac{2}{3}$	$-\frac{1}{2}$	$+\frac{2}{3}$	$-\frac{1}{4}$	+1	$-\frac{1}{2}$	+1	$-\frac{1}{4}$	+1
5	$+\frac{2}{3}$	$-\frac{1}{2}$	$+\frac{2}{3}$	$-\frac{1}{4}$	0	$-\frac{1}{2}$	+1	$-\frac{1}{4}$	+1
6	$+\frac{1}{3}$	$-\frac{1}{4}$	$+\frac{1}{3}$	$-\frac{1}{2}$	0	$-\frac{1}{4}$	0	$-\frac{1}{2}$	+1
7	0	0	0	0	0	-1	0	0	0
8	$-\frac{5}{9}$	$+\frac{5}{12}$	$-\frac{5}{9}$	$-\frac{5}{12}$	0	$+\frac{5}{12}$	0	$-\frac{5}{12}$	0
9	$+\frac{4}{9}$	$-\frac{1}{3}$	$+\frac{4}{9}$	$+\frac{1}{3}$	0	$-\frac{1}{3}$	0	$-\frac{2}{3}$	0

$$\{P\}_{9\times 1} =$$

1	0.0
2	0.0
3	+9.0
4	0.0
5	0.0
6	−18.0
7	0.0
8	0.0
9	0.0

$$\{F\}_{9\times 1} = [A^{-1}]\{P\} =$$

1	−10.0
2	+ 1.5
3	−12.5
4	+15.0
5	+15.0
6	+ 7.5
7	+18.0
8	−12.5
9	+10.0

3.4 Graphical Method of Joints

The *graphical method of joints* is a counterpart of the conventional method of joints. Its use is most suitable in cases where the number of reaction components is three and the configuration consists of a sequence of triangles stacked one upon another. First of all, the reaction components should be computed algebraically. (The graphical method described here applies only to the determination of the internal axial forces.)

The truss of Example 3.1, shown again in Fig. 3.10a, has three reaction components and consists of four triangles which are numbered 1, 2, 3, and 4 beginning from the left. The spaces between external forces and reactions are designated *A-B-C-D-E-A* clockwise around the truss so the force vector *A-B-C-D-E-A* forms a closed polygon in Fig. 3.10b.

Each internal force is named by the two symbols on either side of the bar taken in the clockwise order around a joint; for instance, F_1 is in the direction of 1-*A* in Fig. 3.10b acting on the upper joint in Fig. 3.6a, but it is in the direction of *A*-1 in Fig. 3.10b acting on the lower joint of Fig. 3.10a. Points 1-2-3-4 on the force diagram of Fig. 3.10b are determined in succession; for instance, point 1 is the intersection of *A*-1 parallel to F_1 and *D*-1 parallel to F_4. Thus, one reads *A*-1-*D-E-A* around joint L_0 of the truss diagram and finds the closed polygon of the same name in the force diagram. The axial force diagram, then, is no more than the superposition of all the force polygons for the joint free bodies.

The magnitude of each axial force may be measured from the force diagram, and its kind (tension or compression), determined by observing the direction of

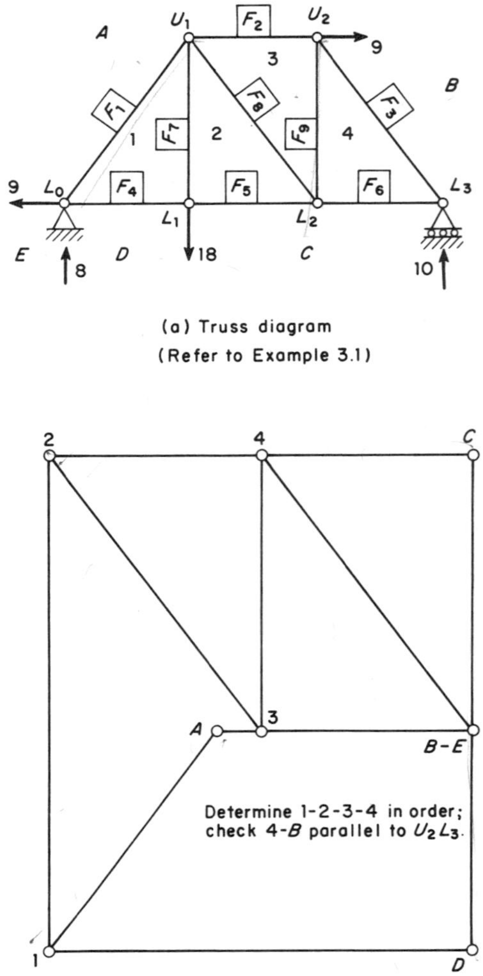

(b) Force diagram

FIG. 3.10. The Graphical Method of Joints.

the force acting on the appropriate joint. For the truss of Fig. 3.10,

$$F_1 = A\text{-}1 \text{ on } L_0 = 1\text{-}A \text{ on } U_1 = 10 \text{ kips compression}$$
$$F_2 = A\text{-}3 \text{ on } U_1 = 3\text{-}A \text{ on } U_2 = 1.5 \text{ kips tension}$$
$$F_3 = B\text{-}4 \text{ on } U_2 = 4\text{-}B \text{ on } L_3 = 12.5 \text{ kips compression}$$
$$F_4 = 1\text{-}D \text{ on } L_0 = D\text{-}1 \text{ on } L_1 = 15 \text{ kips tension}$$
$$F_5 = 2\text{-}C \text{ on } L_1 = C\text{-}2 \text{ on } L_2 = 15 \text{ kips tension}$$
$$F_6 = 4\text{-}C \text{ on } L_2 = C\text{-}4 \text{ on } L_3 = 7.5 \text{ kips tension}$$
$$F_7 = 2\text{-}1 \text{ on } U_1 = 1\text{-}2 \text{ on } L_1 = 18 \text{ kips tension}$$
$$F_8 = 3\text{-}2 \text{ on } U_1 = 2\text{-}3 \text{ on } L_2 = 12.5 \text{ kips compression}$$
$$F_9 = B\text{-}4 \text{ on } U_2 = 4\text{-}B \text{ on } L_3 = 12.5 \text{ kips compression}$$

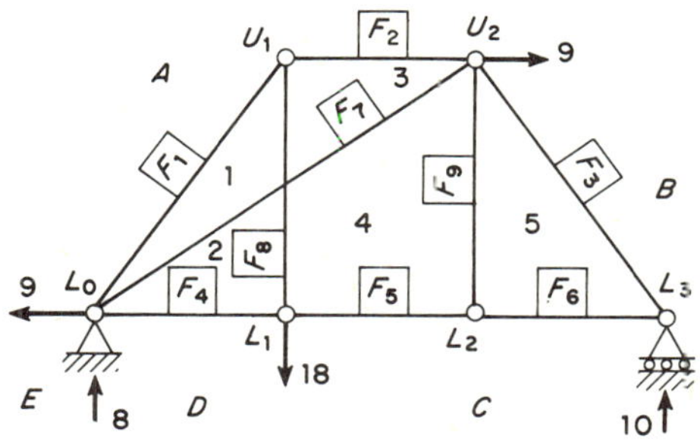

(a) Truss diagram
(Refer to Exercise 3.1)

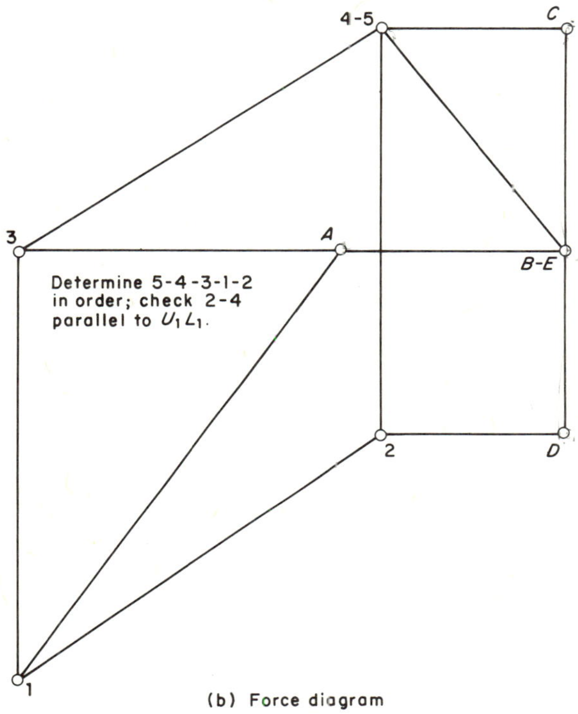

(b) Force diagram

FIG. 3.11. The Graphical Method of Joints.

Note in Fig. 3.11 that in the graphical solution for the truss in Exercise 3.1, the axial force F_8 is equal to 3-1 acting on U_1 or 2-4 acting on L_1, and F_7 is equal to 1-2 acting on L_0 or 4-3 acting on U_2.

EXERCISE 3.7

Analyze the truss of Exercise 3.3 (Fig. 3.6) by the graphical method.

3.5 Deflections by Joint Displacement Equation

When the loads P_1, P_2 and P_3 are gradually applied to an unstressed, undeformed truss ABC as shown in Fig. 3.12a, the stressed, deformed truss will take the shape $A'B'C'$. The displacements A-A', B-B', and C-C' are defined as deflections of the truss; they are functions of the shape of the initial truss, the cross-sectional areas, the modulus of elasticity, and the applied loads.

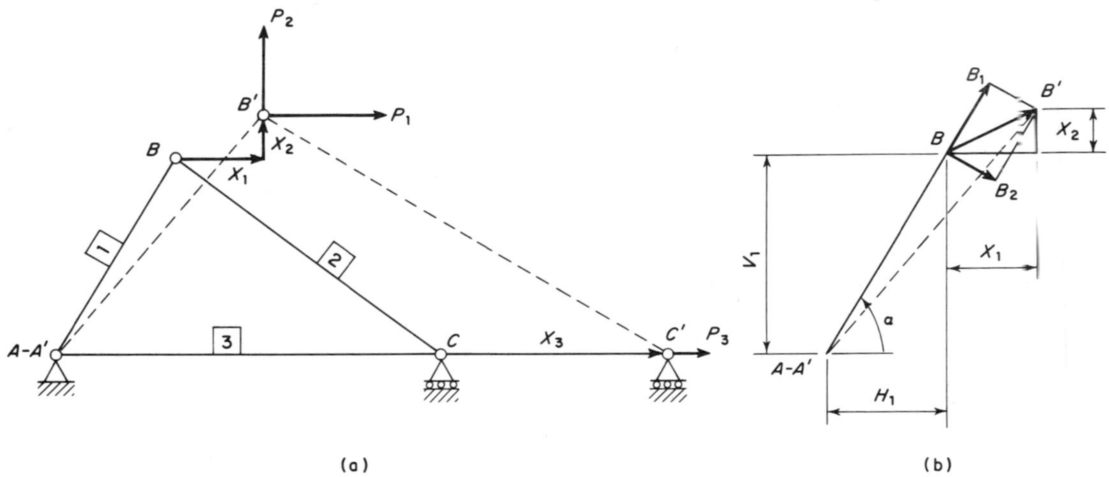

FIG. 3.12. Definition of Truss Deflections.

Consider first the question for determining the loads P_1, P_2, and P_3 if the displacements X_1, X_2, and X_3 are given. This involves the computation of: 1) the elongations of the bars and the axial forces in them; and 2) the forces P_1, P_2, and P_3 to keep the joints B' and C' in equilibrium.

Regarding the elongations of bars, the exact approach would be to say that the elongation e in any bar is equal to

$$e = L' - L_0$$

where L' and L_0 are the deformed and initial lengths, respectively. For instance,

$$e_1(\text{in Fig. 3.12}) = \sqrt{(H_1 + X_1)^2 + (V_1 + X_2)^2} - \sqrt{H_1^2 + V_1^2} \tag{3.5.1}$$

The approximate approach is to resolve the deflection BB' into its axial component BB_1 and transverse component BB_2, and to say that the elongation of bar AB is

$$e_1(\text{in Fig. 3.12}) = BB_1 = X_1 \cos\alpha + X_2 \sin\alpha \tag{3.5.2}$$

When the deflections are small relative to the original dimensions, as they generally are, the second approach is not only sufficiently accurate, but also makes direct computation of truss deflections possible. In the remainder of this textbook, elongations will always be based on this assumption.

Having calculated the elongations, the tensile forces in the bars may be computed by the Hooke's Law as

$$F = \frac{EA}{L_0} e \tag{3.5.3}$$

in which E is the modulus of elasticity, A is the cross-sectional area, and L_0 is the original length of the bar.

Regarding the computation of the joint forces P_1, P_2, and P_3 from the axial forces F_1, F_2, and F_3, one has the choice of using joints B' and C' as the free bodies, or of using the original joints B and C. The first approach is admittedly more exact, but the second approach can give as good results at normal load levels and it will be used in the remainder of this textbook. In fact, in the exact method, which is called *second-order analysis*, the forces F_1-F_2-F_3 and the displacements X_1-X_2-X_3 are so implicit that a direct solution is virtually impossible.

Now consider the use of the joint displacement equation in the computation of the deflections of a statically determinate truss after all reaction components and internal forces have been obtained by a *first-order analysis*. The joint displacement equation is a more general form of Eq. (3.5.2). Referring to Fig. 3.13,

$$e = (X_3 - X_1) \cos \alpha + (X_4 - X_2) \sin \alpha \tag{3.5.4}$$

Note that e as used here represents the elongation in inches or feet.

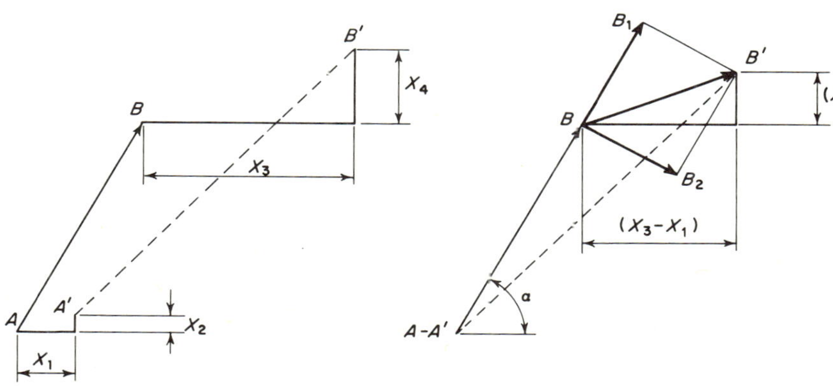

FIG. 3.13. Elongation in Terms of Joint Displacements.

Take for example the forces and elongations for the truss of Example 3.1 as shown in Fig. 3.14a and c. The shape of the deformed truss L_0'-L_1'-L_2'-L_3'-U_1'-U_2' is dictated by the bar elongations and it should be superimposed on the original shape by placing L_0' on L_0, and L_4' on the same elevation as L_4 by means of the hinge and roller supports. However, it is convenient at the beginning to

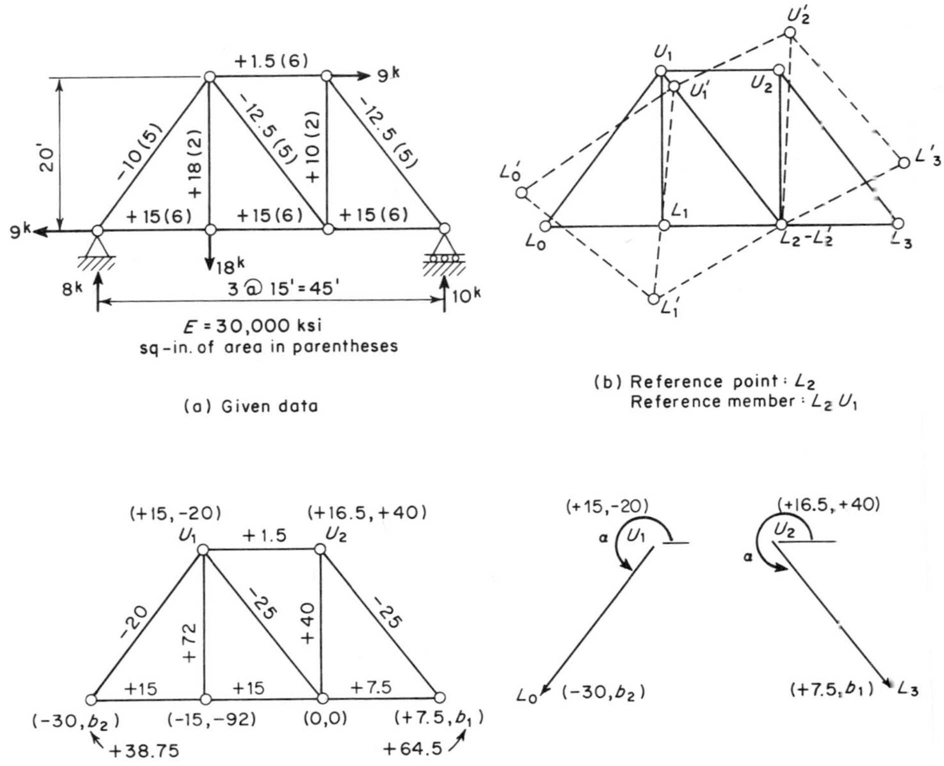

(a) Given data

(b) Reference point: L_2
Reference member: $L_2 U_1$

(c) Elongations and joint displacements in 10^{-3} in.

(d) Angle a in joint displacement equation

+15	−20	+16.5	−40.
+30	−38.75	+30	−38.75
+103/9	−25.75/3	+103/9	−51.50/3
508/9	−202/3	+521.5/9	−47.75/3

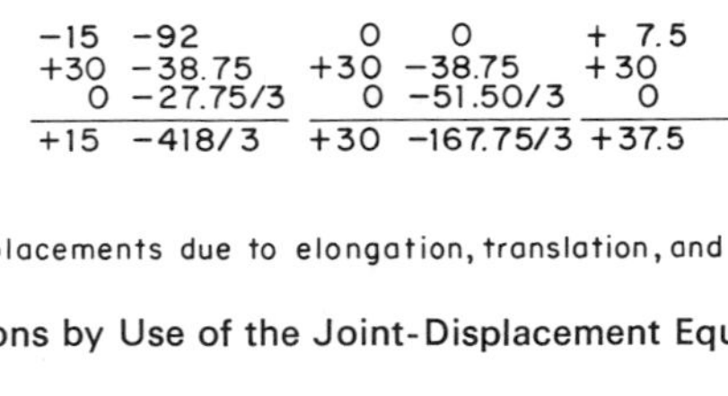

−30	+38.75	−15	−92	0	0	+7.5	+64.50
+30	−38.75	+30	−38.75	+30	−38.75	+30	−38.75
0	0	0	−27.75/3	0	−51.50/3	0	−25.75
0	0	+15	−418/3	+30	−167.75/3	+37.5	0

(e) Joint displacements due to elongation, translation, and rotation

FIG. 3.14. Truss Deflections by Use of the Joint-Displacement Equation.

place, for instance, joint L_2' over joint L_2, and bar L_2'-U_1' over bar L_2-U_1, as shown in Fig. 3.14b. In this orientation, L_2 and L_2U_1 are called *reference point* and *reference member*, respectively.

The horizontal and vertical displacements of all the joints in the reference position shown in Fig. 3.14c are partially obtained by inspection and partially by successively applying the joint displacement equation (3.5.4). In order that U_1L_2 shortens by 25 units, joint U_1 must move 25 units toward L_2, which means 15 units to the right and 20 units downward. Joint U_2 must move 15 units more to the right than joint U_1, or 16.5 units to the right; and it must move upward 40 units more than joint L_2. Joint L_3 moves 7.5 units to the right to accommodate an elongation of 7.5 units in L_2L_3 but its vertical displacement b_1 will be as yet unknown. Joint L_1 moves 15 units to the left and 92 units downward to accommodate an elongation of 15 units in L_1L_2 and 72 units in U_1L_1. Joint L_0 must move the sum of elongations in L_0L_1 and L_1L_2, or 30 units, to the left; but its vertical displacement b_2 is unknown.

The joint displacement equation is applied to members U_2L_3 and U_1L_0, respectively, in order to obtain b_1 and b_2. For member U_2L_3 (Fig. 3.14d),

$$-25 = (+7.5 - 16.5)(+0.6) + (b_1 - 40)(-0.8)$$
$$b_1 = +64.5 \text{ units}$$

For member U_1L_0 (Fig. 3.14d),

$$-20 = (-30 - 15)(-0.6) + (b_2 + 20)(-0.8)$$
$$b_2 = +38.75 \text{ units}$$

The deformed truss now resting in the reference position of Fig. 3.14b or c must be moved as a rigid body through a translation of 30 units to the right and 38.75 units downward so that joint L_0' may fall upon L_0, followed by a clockwise rotation around L_0 or L_0' to bring L_3' (now at 64.50 — 38.75 = 25.75 units above L_3) down to the same level as L_3. The joint displacements due to bar elongations, due to translation, and due to rotation are shown in Fig. 3.14e, together with the final resulting deflections. The horizontal and vertical displacements of any joint, for instance U_2 in Fig. 3.14e, due to a rotation of $\Delta\phi$ around L_0 are:

$$\begin{aligned}\text{Horizontal displacement} &= U_2U_2' \sin\phi = r\,\Delta\phi \sin\phi \\ &= (\text{vertical projection of } r)(\Delta\phi) \\ &= (20 \text{ ft})\left(\frac{25.75 \text{ units}}{45 \text{ ft}}\right) = \frac{103}{9} \text{ units}\end{aligned}$$

$$\begin{aligned}\text{Vertical displacement} &= U_2U_2' \cos\phi = r\,\Delta\phi \cos\phi \\ &= (\text{horizontal projection of } r)(\Delta\phi) \\ &= (30 \text{ ft})\left(\frac{25.75 \text{ units}}{45 \text{ ft}}\right) = \frac{51.50}{3} \text{ units}\end{aligned}$$

The displacements of other joints due to rotation are similarly determined.

The use of the joint displacement equation in finding truss deflections is fully displayed here, not so much to advocate the method itself as to explain the physical meaning of truss deflections.

EXERCISE 3.8

Solve the illustrative example in the previous article by using U_1 and U_1L_1 as the reference point and reference member.

EXERCISE 3.9

Solve the illustrative example in the previous article by using L_0 and L_0U_1 as the reference point and reference member.

3.6 Deflections by Conventional Unit-Load Method

The *conventional unit-load method* may be used to find the joint displacements of particular joints in specific directions, one at a time. Its derivation is based on the broad principle of virtual work to be presented now.

Owing to an arbitrary set of bar elongations e_1 to e_9, the deformed shape of the truss would assume the position of the dashed line relative to the original truss shown in solid line of Fig. 3.15a, resulting in a set of joint displacements X_1 to X_9. The set of bar elongations are said to be compatible with the set of joint displacements; and the state in Fig. 3.15a is called a *compatible state*. Note that the bar elongations may be due to applied loads, temperature changes, or fabrication tolerances.

By a first-order analysis based on the initial geometry of the truss, a set of bar forces, F_1 to F_9, may be obtained for a single external force, such as $P_i = 1$ kip in Fig. 3.15b. The state in Fig. 3.15b is called an *equilibrium state*, for which the free-body diagrams of the six joints are shown in Fig. 3.15c.

The work done by the forces acting on each joint of Fig. 3.15c, in going through the deflection of that joint in Fig. 3.15a, must be zero because the resultant of the forces acting on any joint is equal to zero. The total work obtained by adding up such "virtual" work of all the six joints is still zero; or

$$\begin{aligned} W = {} & \text{work done by } R_1 \text{ to } R_3 + \text{work done by } P_i = 1 \text{ kip} \\ & + \text{work done by nine pairs of } F_1 \text{ to } F_9 = 0 \end{aligned} \tag{3.6.1}$$

The work done by R_1 to R_3 is zero because there is no displacement in the R-directions. The work done by $P_i = 1$ kip is $(+P_iX_i)$, as long as the positive

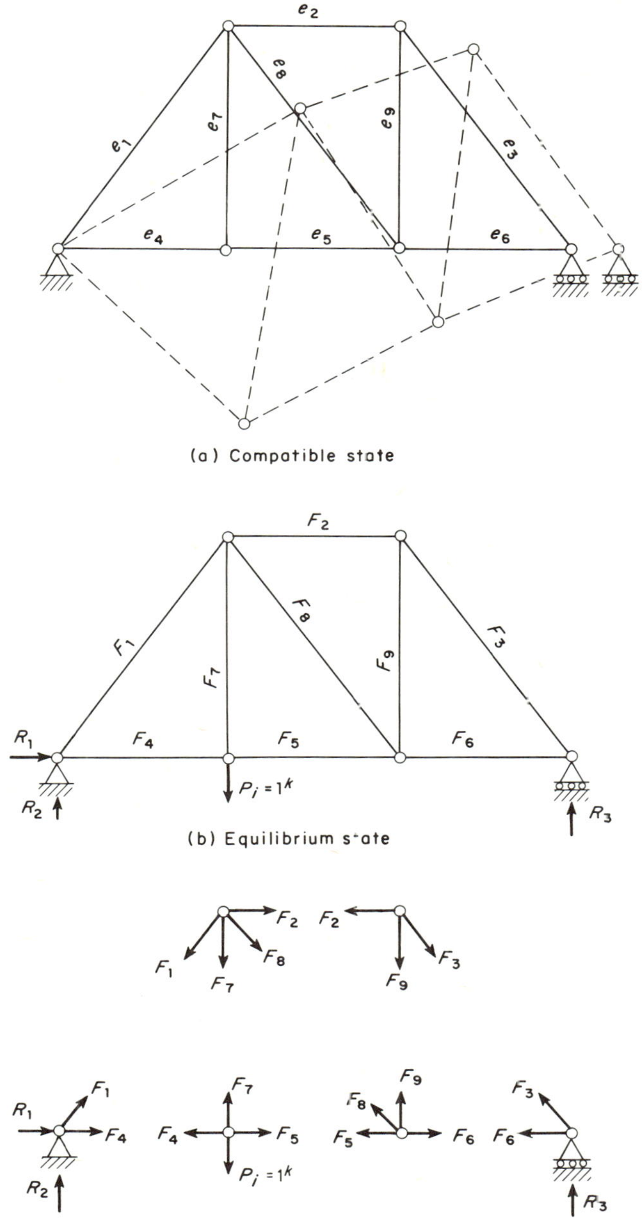

FIG. 3.15. Illustration for the Principle of Virtual Work.

direction of X_i is taken in the same sense as that of P_i. The work done by the force pairs F_1 to F_9 is equal to $(-\Sigma Fe)$ because the two forces in the same pair point toward each other (for tension) in Fig. 3.15c, and the relative displacement between their points of application is away from each other (for elongation) in Fig. 3.15a. Substituting these expressions for virtual work in Eq. (3.6.1), one obtains

$$W = 0 + P_i X_i - \Sigma Fe = 0$$

from which

$$P_i X_i = \Sigma Fe \tag{3.6.2}$$

Equation (3.6.2) in its general form $\Sigma P_i X_i = \Sigma Fe$ (for several P_i-forces) is the Principle of Virtual Work, which states that: "the external work and the

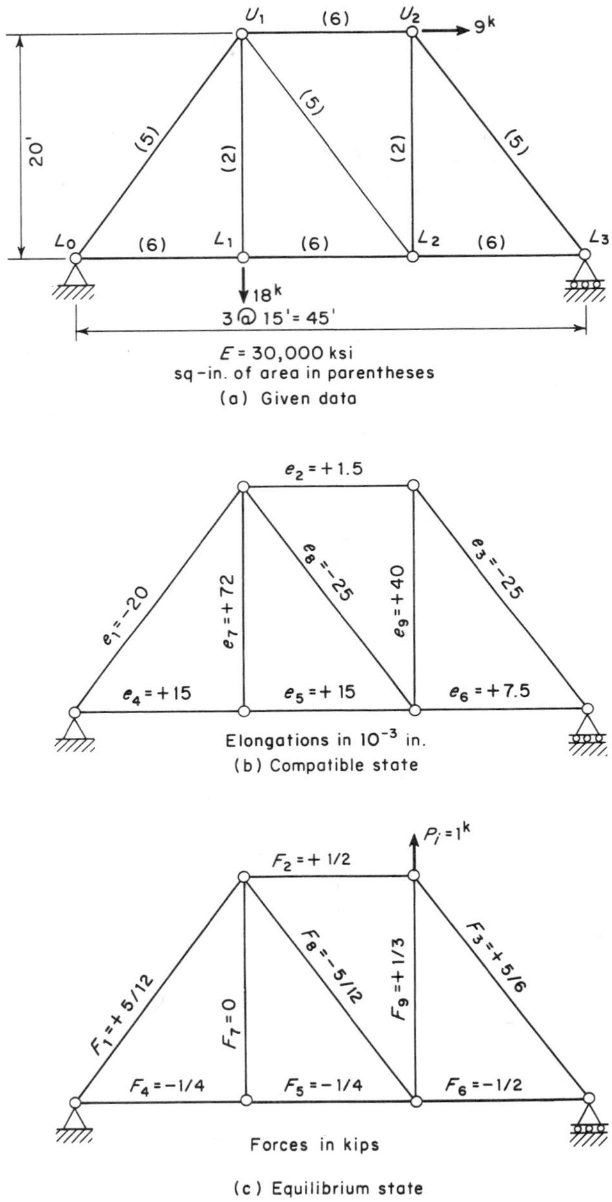

FIG. 3.16. Truss Deflection by the Unit-Load Method.

internal work, each done by the forces in the equilibrium state in going through the deformations of a compatible state, must be equal."

When there is only one unit load in the equilibrium state, the displacement, in the direction of the unit load, in the compatible state may be uniquely found from Eq. (3.6.2); thus the name unit-load method.

Example 3.4. For the bar elongations shown in Fig. 3.16b, find the vertical displacement of joint U_2 by the unit-load method. Compare the results with that in Fig. 3.14.

Solution. Using Fig. 3.16b as the compatible state and Fig. 3.16c as the equilibrium state, and applying the unit-load formula,

$$\begin{aligned} P_i X_i = \Sigma Fe &= \left(+\tfrac{5}{12}\right)(-20) + \left(+\tfrac{1}{2}\right)(+1.5) + \left(+\tfrac{5}{6}\right)(-25) \\ &\quad + \left(-\tfrac{1}{4}\right)(+15) + \left(-\tfrac{1}{4}\right)(+15) - \left(-\tfrac{1}{2}\right)(+7.5) \\ &\quad + (0)(+72) + \left(-\tfrac{5}{12}\right)(-25) + \left(-\tfrac{1}{3}\right)(+40) \\ &= -\frac{47.75}{3} \text{ units} = -\frac{47.75}{3} \times 10^{-3} \text{ in.} \end{aligned}$$

The negative sign shows that the displacement is opposite in direction to the unit-load; or in this case, it is vertically downward.

EXERCISE 3.10

For the bar elongations shown in Fig. 3.16b, find the horizontal displacement of joint U_2 and the vertical displacement of joint L_1 by the unit-load method.

3.7 Deflections by Matrix Unit-Load Method

It has been described in Art. 3.3 that the statics matrix $[A]$ of a truss expresses the external joint forces in terms of the axial forces in the bars; or

$$\{P\}_{NP\times 1} = [A]_{NP\times NF}\{F\}_{NF\times 1}$$

In the event that the truss is statically determinate, $[A]$ is a square matrix; consequently

$$\{F\}_{NF\times 1} = [A^{-1}]_{NF\times NP}\{P\}_{NP\times 1}$$

At this point it is useful to describe one rectangular matrix as being the *transpose* of the other if the ith row of one is the ith column of the other. For instance, if

$$[A] = \begin{bmatrix} 2 & 3 & 4 \\ 5 & 6 & 7 \end{bmatrix} \quad \text{and} \quad [B] = \begin{bmatrix} 2 & 5 \\ 3 & 6 \\ 4 & 7 \end{bmatrix}$$

then,

$$[A] = [B^T] \quad \text{or} \quad [B] = [A^T]$$

Thus the transpose of a square matrix is still a square matrix.

It can be shown that, for a statically determinate truss, the displacements in the directions of all the possible external joint forces may be obtained from the matrix equation

(3.7.1) $$\{X\}_{NP\times 1} = [A^{-1}]^T_{NP\times NF}\{e\}_{NF\times 1}$$

in which the column matrix $\{e\}$ contains the set of arbitrary bar elongations for which the joint displacements $\{X\}$ are to be found.

The general ways of arriving at Eq. (3.7.1) will be deferred for convenience until the next chapter; for the present it will suffice to substantiate this equation by means of Eq. (3.6.2). Since Eq. (3.6.2) is the basic formula for the conventional unit-load method, Eq. (3.7.1) represents the matrix unit-load method.

Equation (3.6.2) states that

(3.7.2) $$\begin{bmatrix} X_1 \\ X_2 \\ \cdot \\ \cdot \\ \cdot \\ X_{NP} \end{bmatrix} = \begin{bmatrix} \sum_{j=1}^{j=NF} (F_{j-1}\, e_j) \\ \sum_{j=1}^{j=NF} (F_{j-2}\, e_j) \\ \cdot \\ \cdot \\ \cdot \\ \sum_{j=1}^{j=NF} (F_{j-NP}\, e_j) \end{bmatrix}$$

in which $F_{j-1}, F_{j-2}, \ldots F_{j-NP}$ are the forces in the jth bar due to a unit load applied in the first, second, . . . NPth direction. Because the columns of the $[A^{-1}]$ matrix show the bar forces due to each unit joint force, the values F_{j-1}, $F_{j-2}, \ldots F_{j-NP}$ are available in the jth row of the $[A^{-1}]$ matrix. Equation (3.7.2) may be written as

$$\begin{aligned} X_1 &= A^{-1}(1, 1) * e(1) + A^{-1}(2, 1) * e(2) + \cdots + A^{-1}(NF, 1) * e(NF) \\ X_2 &= A^{-1}(1, 2) * e(1) + A^{-1}(2, 2) * e(2) + \cdots + A^{-1}(NF, 2) * e(NF) \\ &\cdot\ \cdot\ \cdot\ \cdot\ \cdot\ \cdot \\ X_{NP} &= A^{-1}(1, NP) * e(1) + A^{-1}(2, NP) * e(2) + \cdots + A^{-1}(NF, NP) * e(NF) \end{aligned} \qquad (3.7.3)$$

In longhand matrix form Eq. (3.7.3) is

$$
\begin{bmatrix} X_1 \\ X_2 \\ \cdot \\ \cdot \\ \cdot \\ X_{NP} \end{bmatrix}
=
\begin{bmatrix}
A^{-1}(1,1) & A^{-1}(2,1) & \cdots A^{-1}(NF,1) \\
A^{-1}(1,2) & A^{-1}(2,2) & \cdots A^{-1}(NF,2) \\
\cdot & \cdot & \cdot \\
\cdot & \cdot & \cdot \\
\cdot & \cdot & \cdot \\
A^{-1}(1,NP) & A^{-1}(2,NP) & \cdots A^{-1}(NF,NP)
\end{bmatrix}
\begin{bmatrix} e(1) \\ e(2) \\ \cdot \\ \cdot \\ \cdot \\ e(NF) \end{bmatrix}
\tag{3.7.4}
$$

In compact matrix form Eq. (3.7.4) is

$$\{X\}_{NP\times 1} = [A^{-1}]^T_{NP\times NF}\{e\}_{NF\times 1}$$

which is Eq. (3.7.1).

Example 3.5. For the statically determinate truss shown in Fig. 3.14, obtain the set of joint displacements due to the set of bar elongations by means of the matrix unit-load method. Check the results with those in Fig. 3.14.

Solution. The $[A^{-1}]$ matrix should be computed by inverting the $[A]$ matrix in electronic computation although it is suggested here that the student would compute it directly by the conventional method of joints according to its physical meaning. For this problem it has been so computed in Example 3.3. The $\{e\}$ matrix may be taken from Fig. 3.14c. From Eq. (3.7.1),

$$\{X\} = [A^{-1}]^T\{e\}$$

$$
= \begin{bmatrix}
+\frac{5}{9} & -\frac{1}{3} & -\frac{5}{9} & +\frac{2}{3} & +\frac{2}{3} & +\frac{1}{3} & 0 & -\frac{5}{9} & +\frac{4}{9} \\
+\frac{5}{6} & +\frac{1}{4} & +\frac{5}{12} & -\frac{1}{2} & -\frac{1}{2} & -\frac{1}{4} & 0 & +\frac{5}{12} & -\frac{1}{3} \\
+\frac{5}{9} & +\frac{2}{3} & -\frac{5}{9} & +\frac{2}{3} & +\frac{2}{3} & -\frac{1}{3} & 0 & -\frac{5}{9} & +\frac{4}{9} \\
+\frac{5}{12} & +\frac{1}{2} & +\frac{5}{6} & -\frac{1}{4} & -\frac{1}{4} & -\frac{1}{2} & 0 & -\frac{5}{12} & +\frac{1}{3} \\
0 & 0 & 0 & +1 & 0 & 0 & 0 & 0 & 0 \\
+\frac{5}{6} & +\frac{1}{4} & +\frac{5}{12} & -\frac{1}{2} & -\frac{1}{2} & -\frac{1}{4} & -1 & +\frac{5}{12} & -\frac{1}{3} \\
0 & 0 & 0 & +1 & +1 & 0 & 0 & 0 & 0 \\
+\frac{5}{12} & +\frac{1}{2} & +\frac{5}{6} & -\frac{1}{4} & -\frac{1}{4} & -\frac{1}{2} & 0 & -\frac{5}{12} & -\frac{2}{3} \\
0 & 0 & 0 & +1 & +1 & +1 & 0 & 0 & 0
\end{bmatrix}
*
\begin{bmatrix} -20 \\ +1.5 \\ -25 \\ +15 \\ +15 \\ +7.5 \\ +72 \\ -25 \\ +40 \end{bmatrix}
$$

$$= \begin{bmatrix} +\frac{508}{9} \\ -\frac{202}{3} \\ +\frac{521.5}{9} \\ -\frac{47.75}{3} \\ +15 \\ -\frac{418}{3} \\ +30 \\ -\frac{167.75}{3} \\ +37.5 \end{bmatrix} \times 10^{-3} \text{ in.}$$

EXERCISE 3.11

Compute the horizontal and vertical displacements of all joints of the truss shown in Fig. 3.17 first by successive use of the joint-displacement equation,

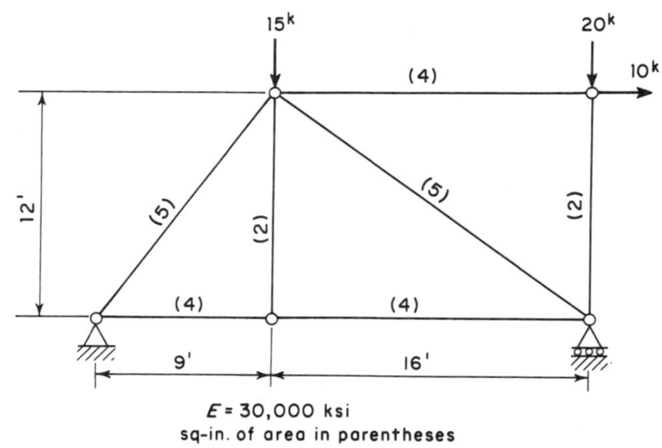

FIG. 3.17. Exercise 3.11.

and then by the matrix unit-load method. In applying the matrix unit-load method, obtain the statics matrix and its inverse independently of each other and verify their correctness by the equation $[A][A^{-1}] = [I]$. Also determine the horizontal displacement of joint U_2 and the vertical displacement of joint L_1 by the conventional unit-load method.

3.8 Concluding Remarks

The statically determinate truss has been singled out from many types of framed structures, and treated first in a substantive way in this chapter, not only for its own importance, but also because of its simplicity in demonstrating fundamental concepts. These concepts include: 1) that the square statics matrix of a statically instable but otherwise statically determinate structure is singular; 2) that the matrix method of joints consists in fact of the solution of a system of linear simultaneous equations equal in number to the number of possible joint forces (later called the degree of freedom in Chapter 4); 3) that the search for the next joint (or section) with two (or three) unknowns, is essential in the conventional method of joints (or sections); 4) that the graphical method of joints has been popular in the past but it is presented here for its own elegance; 5) that the joint-displacement equation is used to emphasize the physical construction of truss deflections in the first-order analysis; and 6) that the Principle of Virtual Work is used to derive the conventional and matrix unit-load method of finding truss deflections.

The suggested exercises are designed to keep the demand upon student time to a minimum. When appropriate, teachers and students are urged to sketch simple trusses and carry through the steps outlined in Exercise 3.11.

Statically Indeterminate Trusses

4

4.1 Degree of Freedom

Typically, a framed structure consists of members connected at joints. A *truss* is a framed structure in which loads are applied only at perfectly smooth pins joining straight bars of uniform cross section. In the previous chapter the symbol *NP* has been used to denote the number of possible directions in which forces may be applied to the joints. It can here be noted that *NP* is also the number of unknown joint displacements; in fact, there is a one-to-one correspondence between joint forces and displacements.

Since the joint displacements represent the ways in which a framed structure may *freely* respond to any external disturbance, the number of unknown joint displacements (or of possible directions of joint forces) will hereafter be called the degree of freedom *NP*.

the reaction R_4 or one of the bars such as U_1U_2, L_1L_2, U_1L_2, or U_2L_2 could have been chosen as the redundant instead of R_3. In Fig. 4.2c, the two redundants may be chosen from two bar forces, or one bar force and one reaction each may be selected, so long as a statically stable and determinate truss remains.

In the *redundant force method*, the redundants must be evaluated at the outset from the compatibility conditions in regard to the relative displacement between the ends of the redundant bar, or to the displacement in the direction of the redundant support. In the former case, the relative displacement must be such that it accommodates the elongation of the redundant bar; and in the latter case, the displacement must be zero at the redundant reaction.

The relative displacement X_i (or increase in the distance) between two joints of a truss may be computed by the conventional unit-load method, provided that a *pair* of unit P_i-forces is applied in the equilibrium condition as shown in Fig. 4.3. Then, applying Eq. (3.6.1) to Fig. 4.3,

(4.3.1) $$\text{Relative displacement } X_i \text{ (apart)} = \Sigma Fe$$

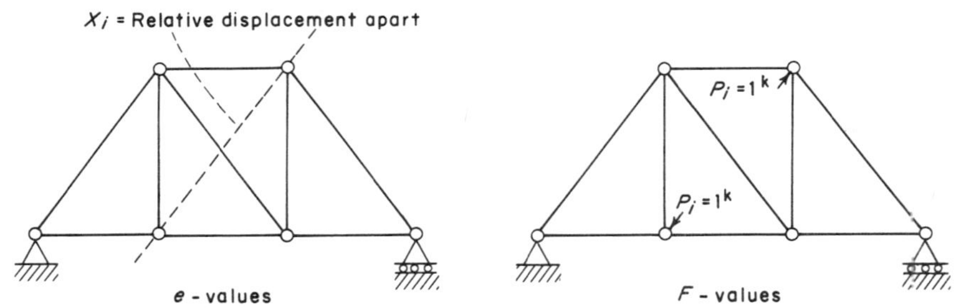

(a) Real compatible condition (b) Dummy equilibrium condition

FIG. 4.3. Relative Displacement by the Conventional Unit-Load Method.

In the event that all the joint displacements in the real compatible condition have been computed as shown in Fig. 4.4, the relative displacement between two typical joints such as L_1 and U_2 becomes

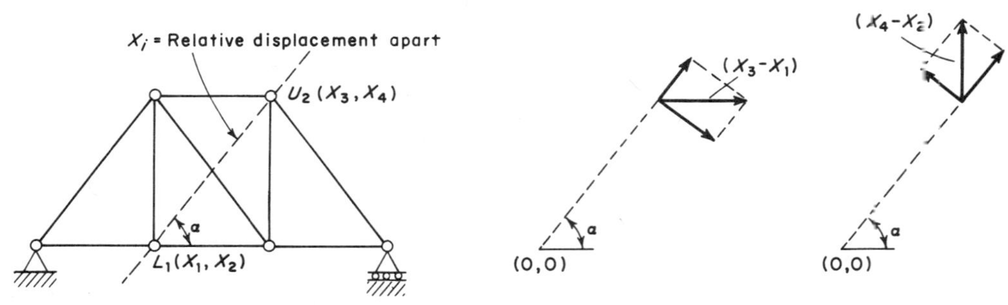

(a) Real e - values (b) Effect of (X_3-X_1) (c) Effect of (X_4-X_2)

FIG. 4.4. Relative Displacement from Horizontal and Vertical Displacements.

(4.3.2) Relative displacement X_i (apart)

$$= (X_3 - X_1)\cos\alpha + (X_4 - X_2)\sin\alpha$$

The above equation can be easily substantiated by taking the components of the relative horizontal and vertical displacements, $(X_3 - X_1)$ and $(X_4 - X_2)$, in the direction of an imaginary line joining L_1 and U_2.

In applying the compatibility conditions, the deflections in the directions of the redundants must be the sum of: 1) those due to the loads and, 2) those due to each redundant force, as applied to the degenerate determinate truss. These conditions yield a system of linear simultaneous equations from which the redundant forces may be solved. Although it is possible to express these simultaneous equations by means of general symbolic expressions, it is deemed sufficient in this introductory text to consider the following two numerical examples.

Example 4.1. Analyze the statically indeterminate truss of Fig. 4.5a by the redundant force method, using the force in bar L_1U_2 as the redundant. Compute all axial forces and joint displacements, and show that all statics and compatibility conditions are satisfied.

Solution. Although the solution as presented in Fig. 4.5 could be either arranged in tabular form or automated by a computer program, the diagrammatic scheme is used here to explain the procedure. The compatibility condition requires that the relative displacement X_i due to the combined action of the applied loads and of the pair of redundant forces must be equal to the elongation of the redundant bar. Applying the conventional unit-load method,

$$\begin{aligned} X_i \text{ (due to applied loads)} &= \Sigma\,(F \text{ in Fig. 4.5d})\,(e \text{ in Fig. 4.5c}) \\ &= (+0.6)(+1.5 - 15) + (+0.8)(+72 + 40) + (-1)(-25) \\ &= +\ 124.5 \text{ units} \end{aligned}$$

$$\begin{aligned} X_i \text{ (due to the pair of } F_x) &= \Sigma\,(F \text{ in Fig. Fig. 4.5d})\,(e \text{ in Fig. 4.5f}) \\ &= 2(+0.6)(-0.6F_x) + 2\,(+0.8)(-3.2F_x) + (-1)(+2F_x) \\ &= -7.84F_x \text{ units} \end{aligned}$$

Since the joint displacements are available in Fig. 4.5c, the relative displacement between joints L_1 and U_2 may be also computed as

$$\begin{aligned} &X_i \text{ (due to applied loads)} \\ &\quad = (+57.94 - 15)(+0.6) - (-15.92 + 139.33)(+0.8) \\ &\quad = +124.5 \text{ units} \end{aligned}$$

Equating the elongation of the redundant bar with the total relative displacement,

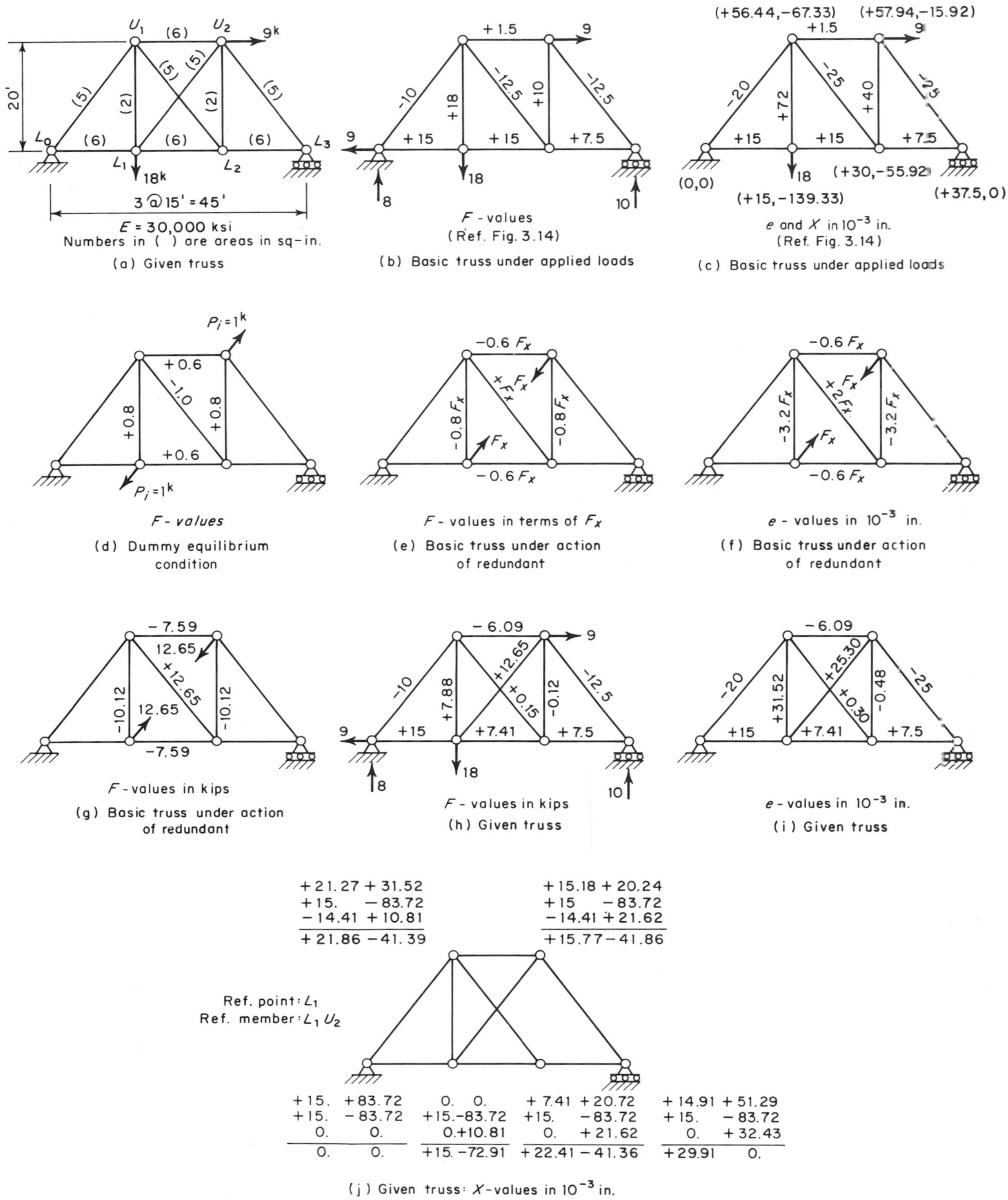

FIG. 4.5. The Redundant Force Method (Example 4.1).

$$\frac{F_x(300)}{(5)(30)} = +124.5 - 7.84F_x$$

$$F_x = +\frac{124.5}{9.84} = +12.65 \text{ kips}$$

The final F-values in the given truss of Fig. 4.5h are the combinations of those in Fig. 4.5b and g. Nine statics checks should be made for the F-values in Fig. 4.5h, one along each degree of freedom.

Joint displacements of the given truss are computed in Fig. 4.5j by means of the joint displacement equation omitting a different bar such as U_2L_2. The compatibility check is obtained from the fact that the vertical displacements of joints U_2 and L_2 in Fig. 4.5j are consistent with the negative elongation in bar U_2L_2. Note that the number of compatibility checks is one, as is the degree of indeterminacy of the given truss.

Example 4.2. Analyze the statically indeterminate truss of Fig. 4.6a by the redundant force method, using the reaction R_3 at L_2 as the redundant. Compute all axial forces and joint displacements, and show that all statics and compatibility conditions are satisfied.

Solution. The compatibility condition requires that the vertical displacement of joint L_2 due to the combined action of the applied loads and the redundant reaction R_3 must be equal to zero. Applying the conventional unit-load method,

Downward vertical displacement of joint L_2 (due to applied loads)

$$\begin{aligned}
&= \Sigma\,(F \text{ in Fig. 4.6d})\,(e \text{ in Fig. 4.6c}) \\
&= (-\tfrac{1}{2})(+1.5) + (+\tfrac{1}{4})(15 + 15) + (-\tfrac{1}{2})(+7.5) + (-\tfrac{5}{12})(-20) \\
&\quad + (+\tfrac{5}{12})(-25) + (-\tfrac{5}{6})(-25) + (-\tfrac{2}{3})(+40) \\
&= +55.92 \text{ units}
\end{aligned}$$

Downward vertical displacement of joint L_2 (due to R_3 upward)

$$\begin{aligned}
&= \Sigma\,(F \text{ in Fig. 4.6d})(e \text{ in Fig. 4.6f}) \\
&= -R_3[2(\tfrac{1}{2})^2 + 2(\tfrac{1}{4})^2 + 2(\tfrac{5}{12})(\tfrac{5}{6}) + (\tfrac{5}{6})(\tfrac{5}{3}) + (\tfrac{2}{3})(\tfrac{8}{3})] \\
&= -4.486R_3 \text{ units}
\end{aligned}$$

Equating the vertical displacement at R_3 with zero,

$$+55.92 - 4.486R_3 = 0$$

$$R_3 = +12.46 \text{ kips}$$

The final F-values in the given truss of Fig. 4.6h are the combinations of those in Fig. 4.6b and g. Eight statics checks should be made for the F-values in Fig. 4.6h, one along each degree of freedom.

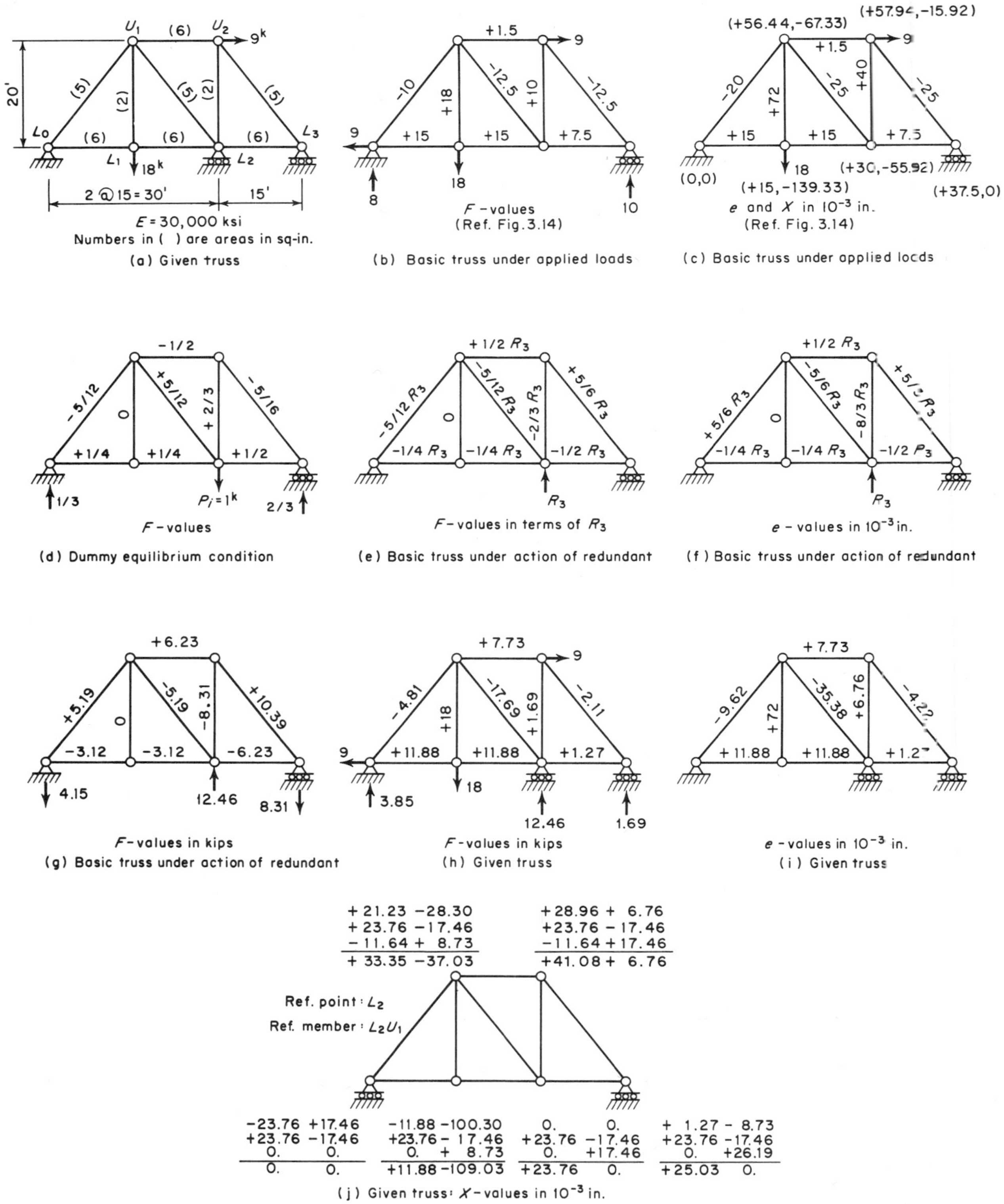

FIG. 4.6. The Redundant Force Method (Example 4.2).

Joint displacements of the given truss are computed in Fig. 4.6j by means of the joint displacement equation. The compatibility check is obtained from the fact that the vertical displacements of joints L_0, L_2, and L_3 are all equal to zero.

EXERCISE 4.1

Analyze the statically indeterminate truss as shown in Fig. 4.7 by the redundant force method, using the force in bar L_1U_2 as the redundant. Compute all axial forces and joint displacements, and show that all statics and compatibility conditions are satisfied.

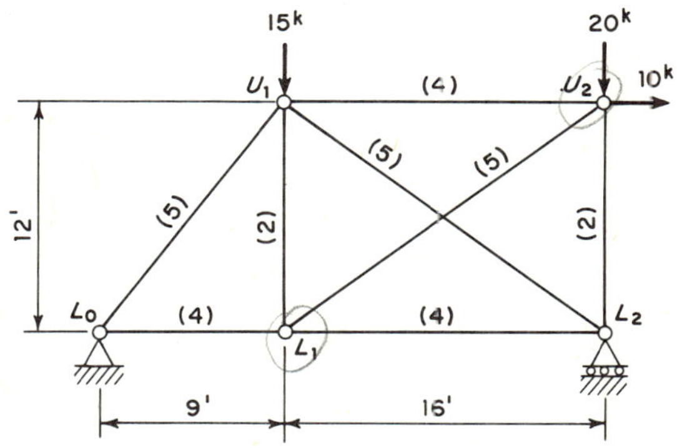

E = 30,000 ksi
Numbers in ( ) are areas in sq-in.

FIG. 4.7. Exercise 4.1.

EXERCISE 4.2

Analyze the statically indeterminate truss as shown in Fig. 4.8 by the redundant force method, using the reaction R_3 at L_1 as the redundant. Compute

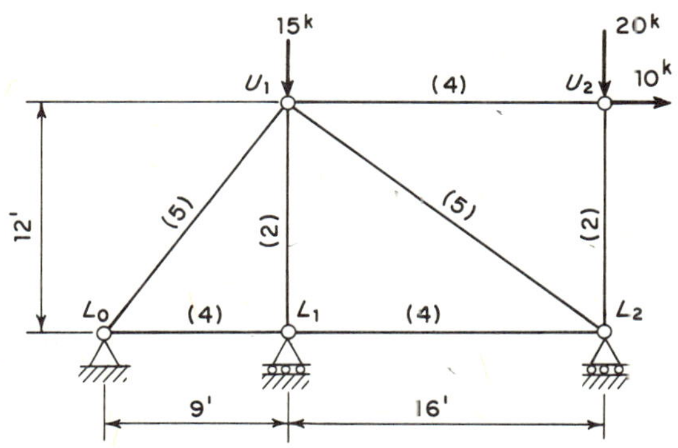

E = 30,000 ksi
Numbers in ( ) are areas in sq-in.

FIG. 4.8. Exercise 4.2.

all axial forces and joint displacements, and show that all statics and compatibility conditions are satisfied.

4.4 The Matrix Displacement Method

In contrast to the redundant force method, which can only be related to statically indeterminate structures, the displacement method applies equally well to both determinate and indeterminate structures. The reason is that in the displacement method the joint displacements are taken as the primary unknowns, and there are always as many equations of equilibrium in the directions of the degree of freedom as there are unknown joint displacements, regardless of the indeterminacy of the structure.

It is most convenient to use matrices to express the linear equations involved in the applications of the displacement method; hence, the name *matrix displacement method.* Two important matrices will be defined: the *internal-force vs. joint-displacement* matrix $[C]$ and the *external-force vs. joint-displacement* matrix $[K]$.

Consider the simple truss of Fig. 4.9. Its degree of freedom NP is four, and the directions of possible joint forces P_1 to P_4 and the corresponding unknown

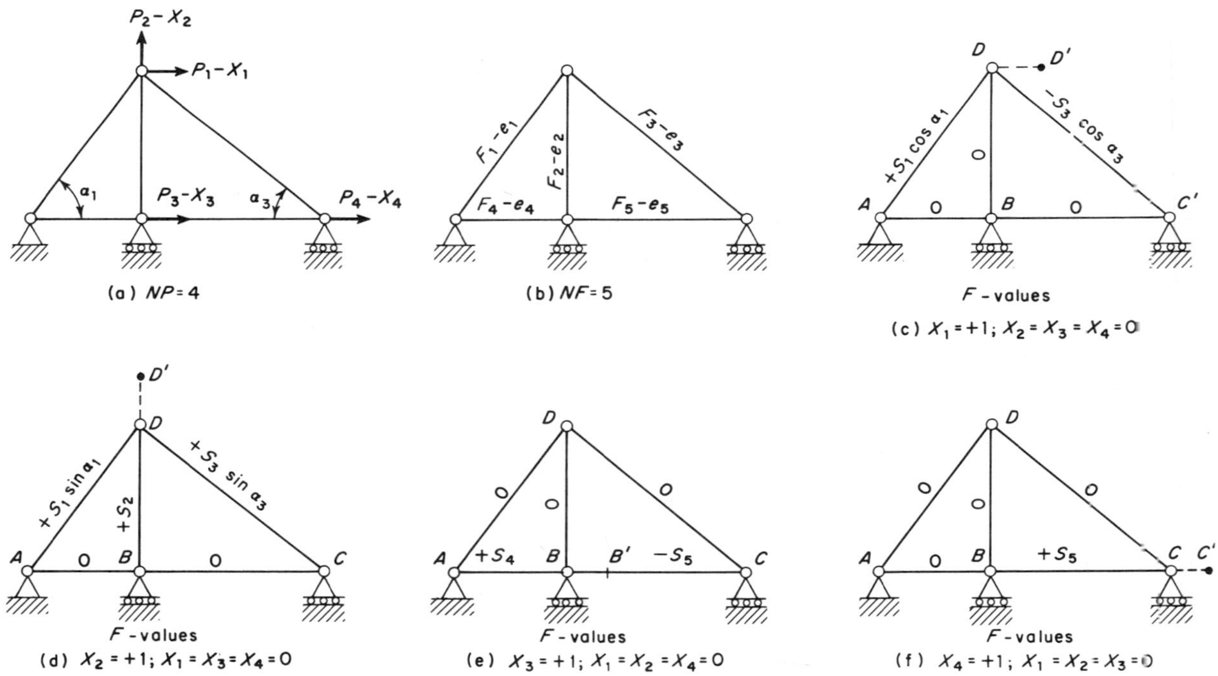

FIG. 4.9. The $[C]$ and $[K]$ Matrices in the Displacement Method.

joint displacements X_1 to X_4 are numbered in succession in Fig. 4.9a. The number of unknown bar forces NF is five; these bar forces and the corresponding elongations are named F_1 to F_5 and e_1 to e_5 in Fig. 4.9b.

The internal-force vs. joint-displacement matrix $[C]$ expresses the bar forces $F(1)$ to $F(NF)$ in terms of the joint displacements $X(1)$ to $X(NP)$. For the truss in Fig. 4.9,

(4.4.1) $[C]_{5\times 4} =$

F \ X	1	2	3	4
1	$+S_1 \cos \alpha_1$	$+S_1 \sin \alpha_1$	0	0
2	0	$+S_2$	0	0
3	$-S_3 \cos \alpha_3$	$+S_3 \sin \alpha_3$	0	0
4	0	0	$+S_4$	0
5	0	0	$-S_5$	$+S_5$

in which $S_1 = EA_1/L_1$ to $S_5 = EA_5/L_5$ are the forces that elongate the respective bars by one unit and thus are measures of the bar stiffnesses. The $[C]$ matrix can be established by columns by noting the forces in the bars of Fig. 4.9c, d, e, and f. Note that only the component of the joint displacement in the bar direction causes a change in the length of the bar.

The external-force vs. joint-displacement matrix $[K]$ expresses the external forces $P(1)$ to $P(NP)$ in terms of the joint displacements $X(1)$ to $X(NP)$. For the truss in Fig. 4.9,

(4.4.2) $[K]_{4\times 4} =$

P \ X	1	2	3	4
1	$+S_1 \cos^2 \alpha_1$ $+S_3 \cos^2 \alpha_3$	$+S_1 \sin \alpha_1 \cos \alpha_1$ $-S_3 \sin \alpha_3 \cos \alpha_3$	0	0
2	$+S_1 \sin \alpha_1 \cos \alpha_1$ $-S_3 \sin \alpha_3 \cos \alpha_3$	$+S_1 \sin^2 \alpha_1 + S_2$ $+S_3 \sin^2 \alpha_3$	0	0
3	0	0	$+S_4+S_5$	$-S_5$
4	0	0	$-S_5$	$+S_5$

Again the $[K]$ matrix can be established by columns by noting the external forces P_1 to P_4 required to maintain equilibrium of the joints in Fig. 4.7c, d, e, and f.

The two important steps in the matrix displacement method are, first, the solution for $\{X\}$ from the equations

$$\{P\} = [K]\{X\} \tag{4.4.3}$$

and then, the computation of $\{F\}$ from the equations

$$\{F\} = [C]\{X\} \tag{4.4.4}$$

A *symmetric matrix* is a square matrix in which the element at (i, j) is equal to the element at (j, i). For instance, the matrix

$$\begin{bmatrix} 7 & 1 & 2 & 3 \\ 1 & 8 & 4 & 5 \\ 2 & 4 & 9 & 6 \\ 3 & 5 & 6 & 10 \end{bmatrix}$$

is symmetric. It will be shown in the next article that the $[K]$ matrix is always symmetric.

Example 4.3. Establish the $[C]$ and $[K]$ matrices of the truss in Fig. 4.10. Using the numerical values of $\{P\}$, $\{X\}$, and $\{F\}$ in Example 4.2, verify numerically the two matrix equations, $\{P\} = [K]\{X\}$ and $\{F\} = [C]\{X\}$, by matrix multiplication.

Solution. The first and second columns of $[C]$ and $[K]$ are established by observing Fig. 4.10f and h. The remaining six columns are similarly determined.

$\{P\} =$

1	0
2	0
3	+9
4	0
5	0
6	−18
7	0
8	0

$\{F\} =$

1	−4.81
2	+7.73
3	−2.11
4	+11.88
5	+11.88
6	+1.27
7	+18.00
8	−17.69
9	+1.69

$\{X\} =$

1	+33.35
2	−37.03
3	+41.08
4	+6.76
5	+11.88
6	−109.03
7	+23.75
8	+25.03

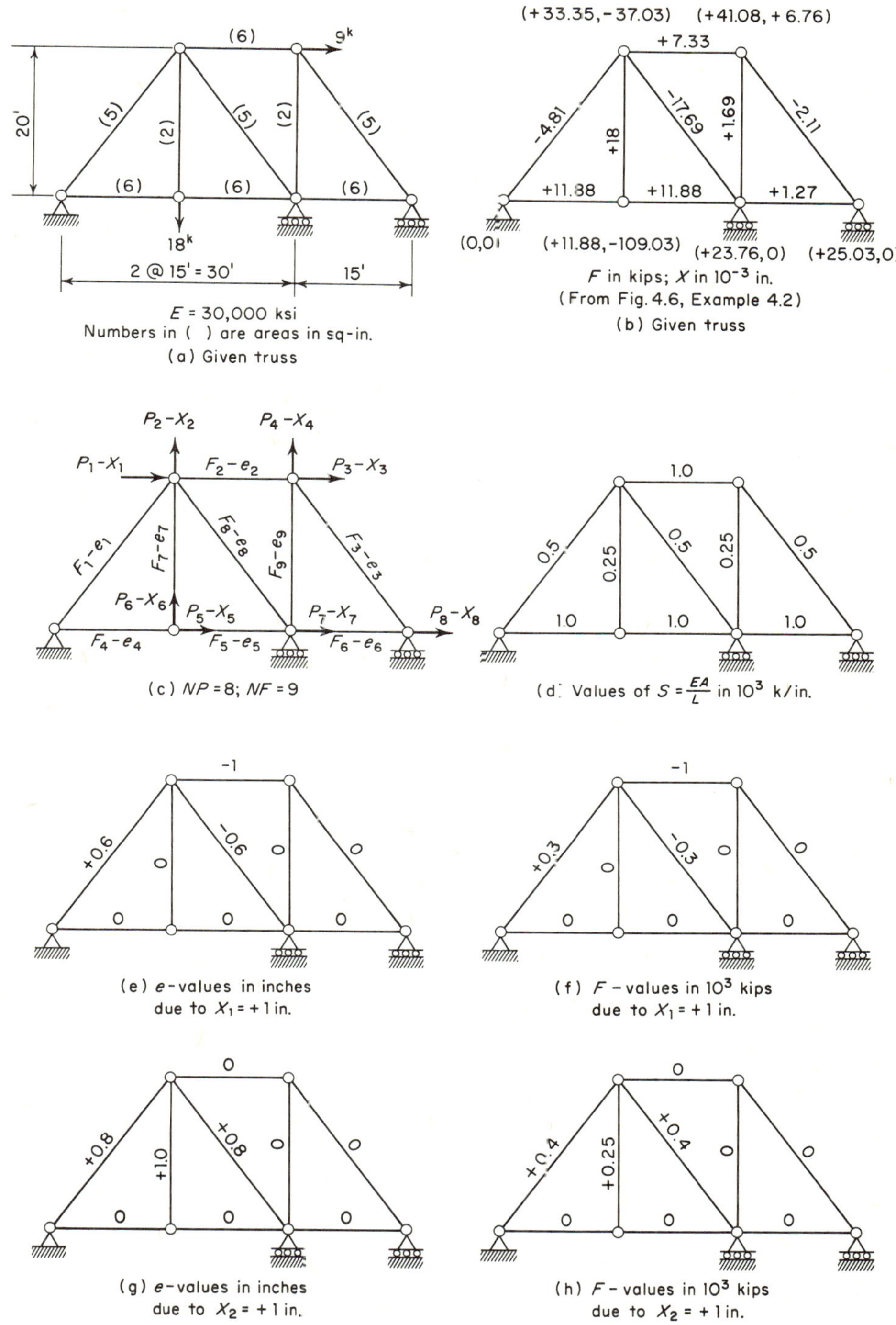

FIG. 4.10. The $[C]$ and $[K]$ Matrices.

F \ X	1	2	3	4	5	6	7	8
1	+0.3	+0.4	0	0	0	0	0	0
2	−1.0	0	+1.0	0	0	0	0	0
3	0	0	−0.3	+0.4	0	0	0	+0.3
4	0	0	0	0	+1.0	0	0	0
5	0	0	0	0	−1.0	0	+1.0	0
6	0	0	0	0	0	0	−1.0	+1.0
7	0	+0.25	0	0	0	−0.25	0	0
8	−0.3	+0.4	0	0	0	0	+0.3	0
9	0	0	0	+0.25	0	0	0	0

$[C] =$ (table above)

P \ X	1	2	3	4	5	6	7	8
1	+1.36	0	−1.00	0	0	0	−0.18	0
2	0	+0.89	0	0	0	−0.25	+0.24	0
3	−1	0	+1.18	−0.24	0	0	0	−0.18
4	0	0	−0.24	+0.59	0	0	0	+0.24
5	0	0	0	0	+2.00	0	−1.00	0
6	0	−0.25	0	0	0	+0.25	0	0
7	−0.18	+0.24	0	0	−1.00	0	+2.18	−1.00
8	0	0	−0.18	+0.24	0	0	−1.00	+1.18

$[K] =$ (table above)

EXERCISE 4.3

Establish the $[C]$ and $[K]$ matrices of the truss in Example 4.1. Using the numerical values of $\{P\}$, $\{X\}$, and $\{F\}$ in that example, verify numerically the equations $\{P\} = [K]\{X\}$ and $\{F\} = [C]\{X\}$.

EXERCISE 4.4

Establish the $[C]$ and $[K]$ matrices of the truss in Exercise 4.1 (Fig. 4.7). Using the numerical values of $\{P\}$, $\{X\}$, and $\{F\}$ in that exercise, verify numerically the equations $\{P\} = [K]\{X\}$ and $\{F\} = [C]\{X\}$.

EXERCISE 4.5

Establish the $[C]$ and $[K]$ matrices of the truss in Exercise 4.2 (Fig. 4.8). Using the numerical values of $\{P\}$, $\{X\}$, and $\{F\}$ in that exercise, verify numerically the equations $\{P\} = [K]\{X\}$ and $\{F\} = [C]\{X\}$.

4.5 Decomposition of Matrices $[C]$ and $[K]$

In the previous article, the internal-force vs. joint-displacement matrix $[C]$ and the external-force vs. joint-displacement matrix $[K]$ have been defined. In the numerical examples it was shown that these two matrices can be established directly by columns. A statement was also made that the $[K]$ matrix is always symmetric.

In this article it will be shown that the matrix $[C]$ is the product of a member stiffness matrix $[S]$ and a deformation matrix $[B]$; or

$$[C] = [S][B] \tag{4.5.1}$$

and that the matrix $[K]$ is the triple product of a statics matrix $[A]$, and the matrices $[S]$ and $[B]$; or

$$[K] = [A][S][B] \tag{4.5.2}$$

Furthermore, it can be shown that the deformation matrix $[B]$ is the transpose of the statics matrix $[A]$; or

$$[B] = [A^T] \tag{4.5.3}$$

For the simple truss of Fig. 4.11, which has been considered before in Fig. 4.9, the four equations of statics along the four degrees of freedom may be obtained by referring to the free-body diagrams of Fig. 4.11c; thus

$$\begin{aligned} P_1 &= +F_1 \cos \alpha_1 - F_3 \cos \alpha_3 \\ P_2 &= +F_1 \sin \alpha_1 + F_2 + F_3 \sin \alpha_3 \\ P_3 &= +F_4 - F_5 \\ P_4 &= +F_3 \cos \alpha_3 + F_5 \end{aligned} \tag{4.5.4a}$$

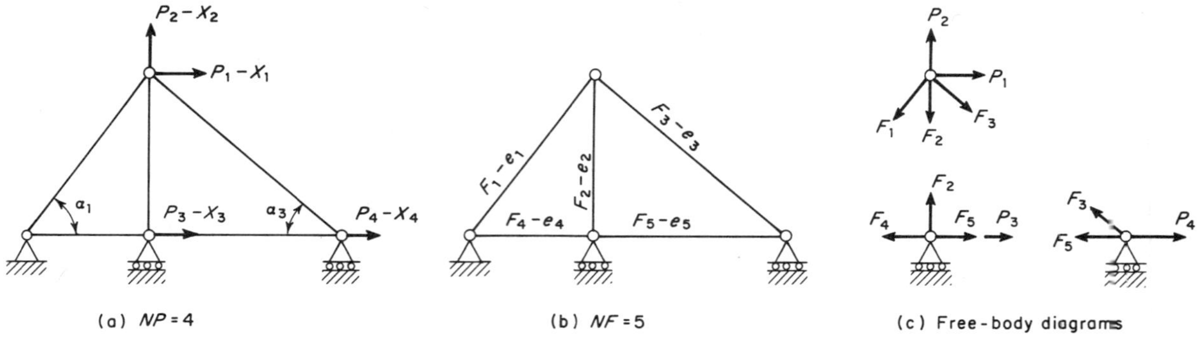

FIG. 4.11. Decomposition of Matrices $[C]$ and $[K]$.

Expressed in matrix form, Eq. (4.5.4a) becomes

(4.5.4b) $$\{P\}_{4\times 1} = [A]_{4\times 5}\{F_{5\times 1}\}$$

in which

$[A]_{4\times 5} =$

P \ F	1	2	3	4	5
1	$+\cos\alpha_1$	0	$-\cos\alpha_3$	0	0
2	$+\sin\alpha_1$	$+1$	$+\sin\alpha_3$	0	0
3	0	0	0	$+1$	-1
4	0	0	$+\cos\alpha_3$	0	$+1$

Since the $[A]$ matrix, as illustrated by Eq. (4.5.4), is based on equations of statics, it is given the name *statics matrix.*

Now examine how the elongation of each bar may be affected by each joint displacement. On the basis of first-order analysis wherein only the component of the joint displacement along the bar length changes its length,

(4.5.5a)
$$\begin{aligned} e_1 &= +X_1 \cos\alpha_1 + X_2 \sin\alpha_1 \\ e_2 &= +X_2 \\ e_3 &= -X_1 \cos\alpha_3 + X_2 \sin\alpha_3 + X_4 \cos\alpha_3 \\ e_4 &= +X_3 \\ e_5 &= -X_3 + X_4 \end{aligned}$$

Expressed in matrix form, Eq. (4.5.5a) becomes

(4.5.5b) $$\{e\}_{5\times 1} = [B]_{5\times 4}\{X\}_{4\times 1}$$

in which

$[B]_{5\times4} =$

e \ X	1	2	3	4
1	$+\cos\alpha_1$	$+\sin\alpha_1$	0	0
2	0	$+1$	0	0
3	$+\cos\alpha_3$	$+\sin\alpha_3$	0	$+\cos\alpha_3$
4	0	0	$+1$	0
5	0	0	-1	$+1$

Since the $[B]$ matrix is entirely based on deformation relationships, it is given the name *deformation matrix.*

Examination of Eqs. (4.5.4b) and (4.5.5b) reveals that the deformation matrix $[B]$ is the transpose of the statics matrix $[A]$, and vice versa. Proof for the transposition relationship may be made by use of the Principle of Virtual Work, which states that the external work and the internal work, each done by the forces in the equilibrium state in going through the deformations of a compatible state, must be equal. Taking the first situation of Fig. 4.12a as the equilibrium state and the second situation of Fig. 4.12b as the compatible state,

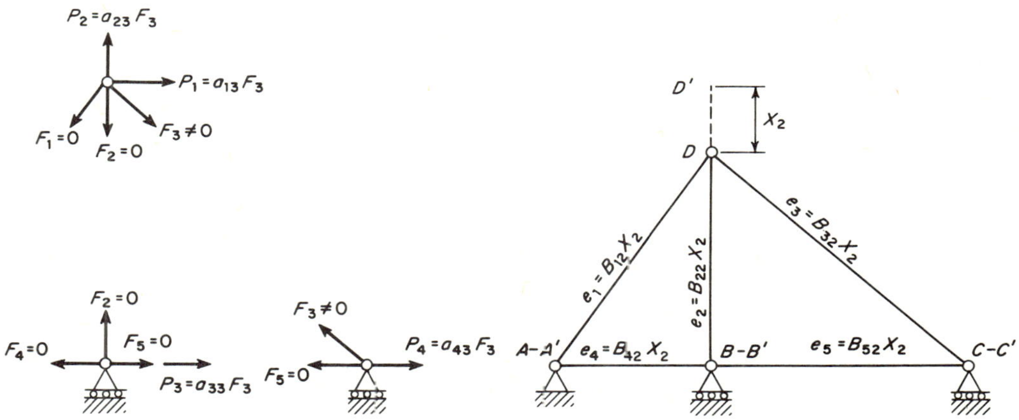

(a) First situation; Equilibrium state ($F_1 = F_2 = F_4 = F_5 = 0; F_3 \neq 0$)

(b) Second situation; Compatible state ($X_1 = X_3 = X_4 = 0$; $X_2 \neq 0$)

FIG. 4.12. Transposition Relationship Between $[A]$ and $[B]$.

$$\Sigma PX = \Sigma Fe$$

$$(a_{23}F_3)(X_2) = (F_3)(b_{32}X_2)$$

$$a_{23} = b_{32}$$

In general,

$$a_{ij} = b_{ij}$$

Thus,

$$[A] = [B^T]; \quad \text{or} \quad [B] = [A^T]$$

which is Eq. (4.5.3).

Applying Hooke's Law, the relationships between bar forces and elongations are

(4.5.6a) $$F_1 = \frac{EA_1}{L_1}e_1; \quad F_2 = \frac{EA_2}{L_2}e_2; \quad \ldots \quad F_5 = \frac{EA_5}{L_5}e_5$$

Expressed in matrix form, Eq. (4.5.6a) becomes

(4.5.6b) $$F_{5\times 1} = [S]_{5\times 5}\,\{e\}_{5\times 1}$$

in which

$[S]_{5\times 5} =$

F \ e	1	2	3	4	5
1	$\frac{EA_1}{L_1}$				
2		$\frac{EA_2}{L_2}$			
3			$\frac{EA_3}{L_3}$		
4				$\frac{EA_4}{L_4}$	
5					$\frac{EA_5}{L_5}$

Since the $[S]$ matrix contains the respective forces required to elongate the bars by a unit length, it is given the name *member stiffness matrix*.

Combining Eqs. (4.5.5b) and (4.5.6b),

(4.5.7) $$\{F\} = [S]\{e\} = [S][B]\{X\}$$

$[B] =$

e \ X	1	2	3	4	5	6	7	8
1	+.6	+.8						
2	−1.		+1.					
3			−.6	+.8				+.6
4					+1.			
5					−1.		+1.	
6							−1.	+1.
7		+1.				−1.		
8	−.6	+.8					+.6	
9				+1.				

$[S] =$ 10^3 k/in.

F \ e	1	2	3	4	5	6	7	8	9
1	0.5								
1		1.0							
3			0.5						
4				1.0					
5					1.0				
6						1.0			
7							0.25		
8								0.5	
9									0.25

$[C] =$ 10^3 k/in.

F \ X	1	2	3	4	5	6	7	8
1	+0.3	+0.4						
2	−1.0		+1.0					
3			−0.3	+0.4				+0.3
4					+1.0			
5					−1.0		+1.0	
6							−1.0	+1.0
7		+0.25				−0.25		
8	−0.3	+0.4					+0.3	
9				+0.25				

$[A] =$

P \ F	1	2	3	4	5	6	7	8	9
1	+0.6	−1.0						−0.6	
2	+0.8						+1.0	+0.8	
3		+1.0	−0.6						
4			+0.8						+1.0
5				+1.0	−1.0				
6							−1.0		
7					+1.0	−1.0		+0.6	
8			+0.6			+1.0			

$[K] =$ 10^3 k/in.

P \ X	1	2	3	4	5	6	7	8
1	+1.36	0.0	−1.0	0.0	0.0	0.0	−0.18	0.0
2	0.0	+0.89	0.0	0.0	0.0	−0.25	+0.24	0.0
3	−1.0	0.0	+1.18	−0.24	0.0	0.0	0.0	−0.18
4	0.0	0.0	−0.24	+0.59	0.0	0.0	0.0	+0.24
5	0.0	0.0	0.0	0.0	+2.0	0.0	−1.0	0.0
6	0.0	−0.25	0.0	0.0	0.0	+0.25	0.0	0.0
7	−0.18	+0.24	0.0	0.0	−1.0	0.0	+2.18	−1.0
8	0.0	0.0	−0.18	+0.24	0.0	0.0	−1.0	+1.18

FIG. 4.13. Solution of Example 4.4.

and, combining Eqs. (4.5.4b) and (4.5.7),

$$\{P\} = [A]\{F\} = [A][S][B]\{X\} \tag{4.5.8}$$

Since the matrix equations $\{F\} = [C]\{X\}$ and $\{P\} = [K]\{X\}$ have been defined in the preceding article, Eqs. (4.5.1) and (4.5.2) are now substantiated.

In computer analysis it is far more convenient to input the matrices $[A]$, $[S]$, and $\{P\}$ and let the computer yield the output matrices $\{X\}$ and $\{F\}$ by the following operations:

1) $\{X\} = [ASA^T]^{-1}\{P\}$
2) $\{F\} = [SA^T]\{X\}$

Based on a common formula in matrix algebra, which states that

$$[AB]^T = [B^T][A^T]$$

the matrix $[K]$ can be shown to be symmetric by the following:

Given $[A]$, $[S] = [S^T]$, $[K] = [ASA^T]$

To Prove $[K] = [K^T]$

Proof Right side $= [K^T] = [ASA^T]^T = [SA^T]^T[A]^T = [A][S]^T[A^T]$
$= [A][S][A^T] = [K] =$ left side

Example 4.4. Establish the $[A]$, $[S]$, and $[B]$ matrices of the truss in Fig. 4.10, and verify the relationship $[B] = [A^T]$. Compute $[C] = [S][B]$ and $[K] = [A][S][B]$ and note that the same results have been obtained previously in Example 4.3.

Solution. Based on the P–X and F–e diagrams in Fig. 4.10, the necessary tabular display for the solution is shown in Fig. 4.13.

EXERCISE 4.6

Rather than establish the $[C]$ and $[K]$ matrices directly by columns according to their own definitions, compute these two matrices for Exercise 4.5 by the equations $[C] = [S][B]$ and $[K] = [A][S][B]$.

4.6 Concluding Remarks

The essential features of the force and displacement methods of structural analysis have been described by means of trusses.

The force method should rightfully be called by its full name "redundant-force" method, because the redundant forces in indeterminate trusses are the primary unknowns.

In the precomputer age the displacement method was not as much used in truss analysis as in continuous-beam or rigid-frame analysis, chiefly because of the large number of simultaneous equations involved in solving for $\{X\}$ from $\{X\} = [ASA^T]^{-1}\{P\}$.

The force method is directly related to the Least-Work Theorem for statically indeterminate structures. This theorem, although applicable as well to indeterminate trusses, will be developed in connection with continuous-beam analysis in Chapter 5.

Statically Determinate Beams

5

5.1 Definition of a Statically Determinate Beam

A *statically determinate beam* is one which can be completely analyzed by the equations of statics. Simple, cantilever, and overhanging beams as shown in Fig. 5.1a, b, and c are three common types of statically determinate beams. Since there exist only two independent equations of statics when the whole beam is taken as a free body, each of these three types of beams contains only two unknown reactions, R_1 and R_2. However, the beam of Fig. 5.1d is still statically determinate, even though it has four unknown reactions, because the existence of two internal hinges yields two additional conditions of statics which require that the bending moment at each internal hinge be zero.

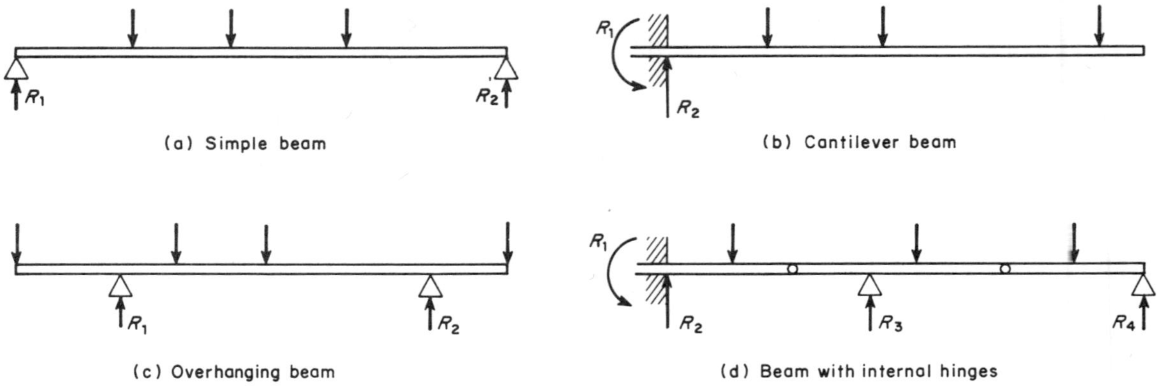

FIG. 5.1. Statically Determinate Beams.

5.2 Shears and Bending Moments in Beams

When a beam is separated into two parts at a section, the action of either part on the other consists of a shear force V and a bending moment M, positive in sign for the directions shown in Fig. 5.2b. The shear at a section is positive if the left side of the beam tends to move upward when the section is cut, as shown in Fig. 5.2c; the bending moment at a section is positive if the beam is concave at the top due to its action. That the signs in Fig. 5.2b, c, and d are consistent may be visualized from the condition that the right part needs to act on the left part by a downward force V to keep the left part from shifting upward relative to the right part, and that the moments M-M are produced by tensile and compressive stresses acting, respectively, on the lower and upper portions of the section.

By considering the equilibrium of the left free body in Fig. 5.1b, the shear

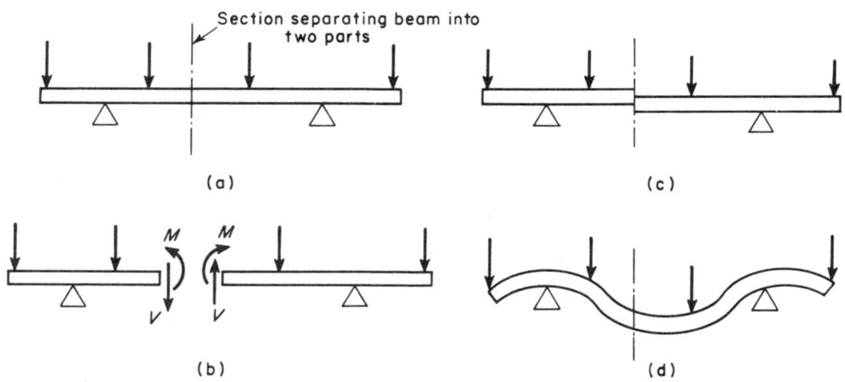

FIG. 5.2. The Shear and Bending Moment at a Section.

V is the sum of all *upward* forces between the left end and the section; and the bending moment M is the sum of all *clockwise* moments about the section. If the right free body is used, however, the shear V is the sum of all *downward* forces between the right end and the section, but the bending moment M is the sum of all *counterclockwise* moments about the section.

Information on the shear and bending moment at a section is needed in the design of the strength of the beam at that section.

Example 5.1. For the beam in Fig. 5.3, compute the reactions R_1 and R_2, and the values of V and M at a section 8 ft to the right of the left support.

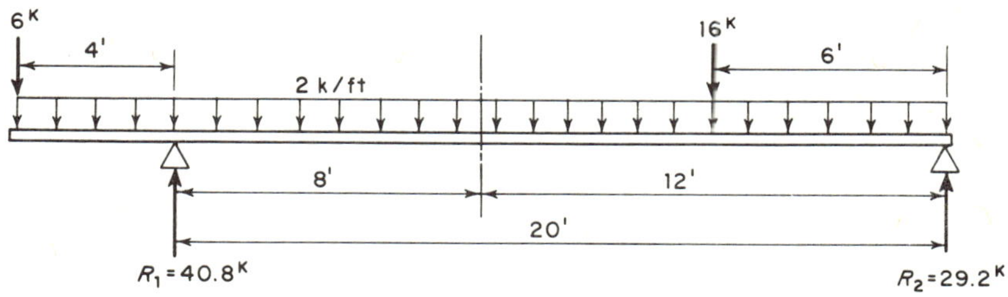

FIG. 5.3. Beam of Example 5.1.

Solution. (a) Reactions R_1 and R_2. Taking moments about R_2,

$$20R_1 = 16(6) + 6(24) + 48(12); \qquad R_1 = 40.8 \text{ kips}$$

Taking moments about R_1,

$$20R_2 = 16(14) - 6(4) + 48(8); \qquad R_2 = 29.2 \text{ kips}$$

Checking by $\Sigma F_y = 0$,

$$40.8 + 29.2 = 6 + 16 + 48; \qquad 70 = 70$$

(b) V and M at 8 ft to right of R_1. Using the left free body,

$$V = +40.8 - 6 - 24 = +10.8 \text{ kips}$$
$$M = 40.8(8) - 6(12) - 24(6) = +110.4 \text{ ft-kips}$$

Checking by using the right free body,

$$V = +16 + 24 - 29.2 = +10.8 \text{ kips}$$
$$M = 29.2(12) - 16(6) - 24(6) = +110.4 \text{ ft-kips}$$

Example 5.2. For the beam in Fig. 5.4, compute the reactions R_1 to R_4, and the bending moment M at R_3.

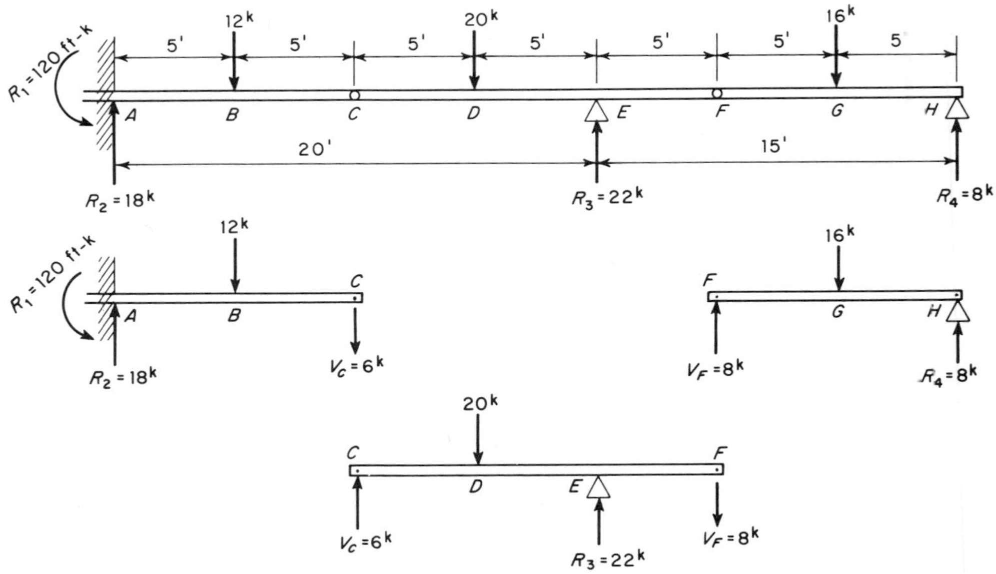

FIG. 5.4. Beam of Example 5.2.

Solution

(a) *Reactions R_1 to R_4.* Taking FGH as a free body,

$$V_F = 8 \text{ kips}; \qquad R_4 = 8 \text{ kips}$$

Taking $CDEF$ as a free body,

$$V_C = 6 \text{ kips}; \qquad R_3 = 22 \text{ kips}$$

Taking ABC as a free body,

$$R_1 = 120 \text{ ft-kips}; \qquad R_2 = 18 \text{ kips}$$

Checking by taking the whole beam as a free body,

$$18 + 22 + 8 = 12 + 20 + 16; \qquad 48 = 48$$

$$\Sigma M_A = 120 + 22(20) + 8(35) - 12(5) - 20(15) - 16(30) = 840 - 840 = 0$$

(b) *Bending moment M at R_3.* Using the left free body,

$$M \text{ at } R_3 = -120 + 18(20) - 12(15) - 20(5) = -40 \text{ ft-kips}$$

Using the right free body,

$$M \text{ at } R_3 = 8(15) - 16(10) = -40 \text{ ft-kips}$$

EXERCISE 5.1

For the given beam shown in Fig. 5.5 compute the reactions R_1 and R_2, and the values of V and M at a section 10 ft to the right of the left support.

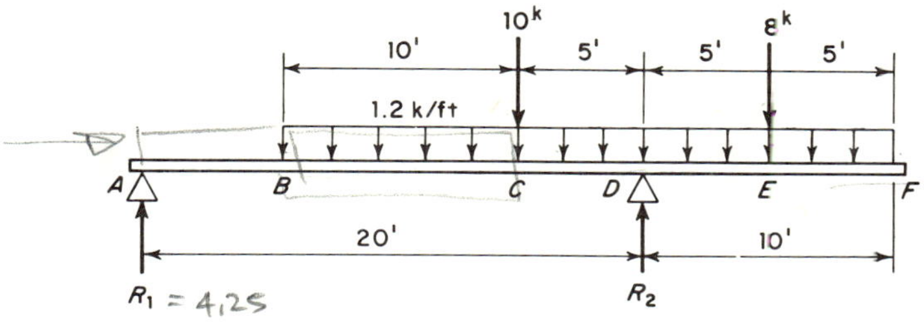

FIG. 5.5. Exercise 5.1.

EXERCISE 5.2

For the given beam shown in Fig. 5.6 compute the reactions R_1 and R_2, and the values of V and M at a section 10 ft from the left end.

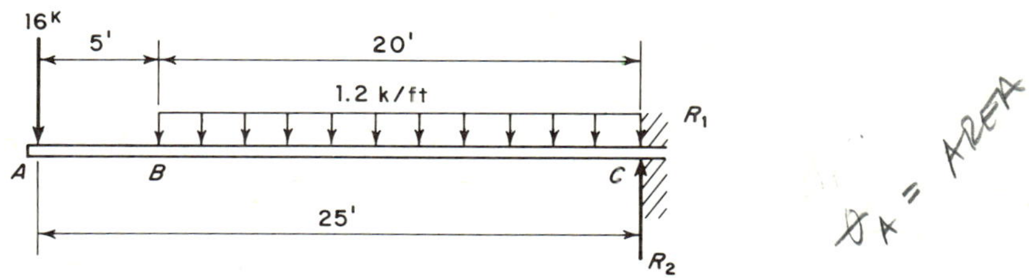

FIG. 5.6. Exercise 5.2.

EXERCISE 5.3

For the given beam shown in Fig. 5.7 compute the reactions R_1 to R_4, and the values of V and M at a section halfway between R_2 and R_3.

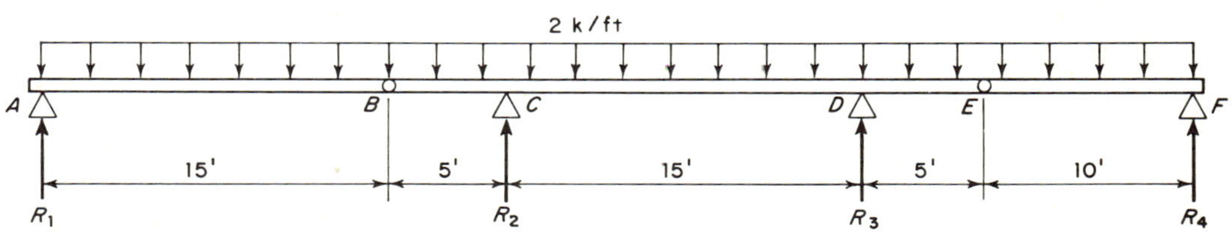

FIG. 5.7. Exercise 5.3.

5.3 Relationships between Load, Shear, and Moment Equations

Sections of discontinuity usually exist along the length of a beam; these are sections on which a concentrated load or reaction acts, or at which the distributed load changes intensity or pattern. The load, shear, and moment at any point between two adjacent sections of discontinuity may be expressed in terms of one independent variable which represents the distance of the point from an arbitrarily chosen reference point. These expressions are called *load, shear,* and *moment equations.*

Relationships exist between the equations for the distributed load intensity q, the shear V, and the moment M at a section in a beam. Applying the two equations of statics to the free-body diagram of a differential length dx shown in Fig. 5.8,

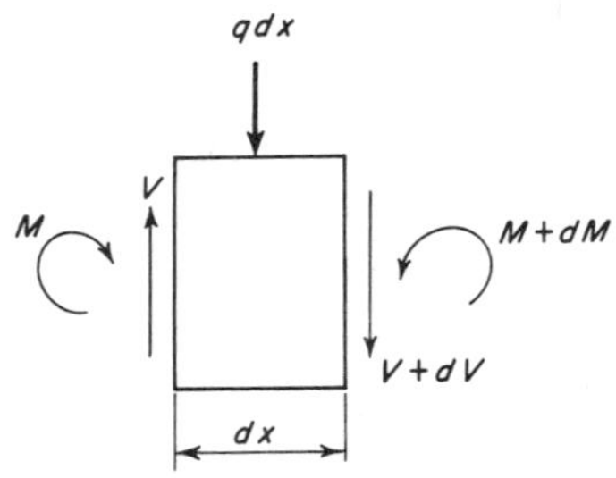

FIG. 5.8. The Free-Body Diagram of a Differential Length *dx*.

$$\frac{dV}{dx} = -q \tag{5.3.1}$$

and,

$$\frac{dM}{dx} = +V \tag{5.3.2}$$

Combination of Eqs. (5.3.1) and (5.3.2) gives

$$\frac{d^2M}{dx^2} = -q \tag{5.3.3}$$

Example 5.3. For the beam of Example 5.1 (shown again in Fig. 5.9), obtain equations for q, V, and M at a section inside BC using its distance x

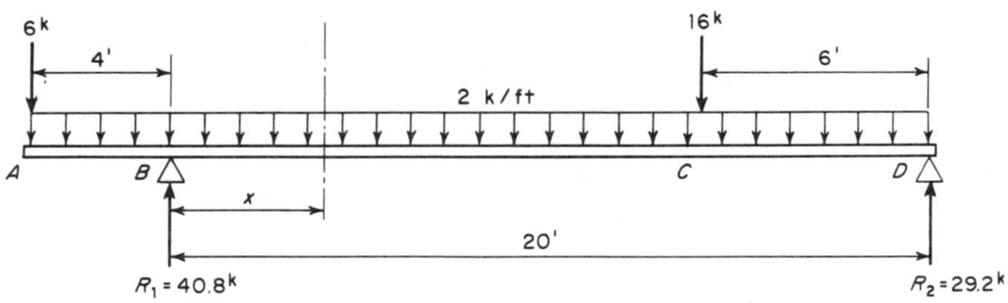

FIG. 5.9. Beam of Example 5.3.

from B as the independent variable, first by using the left free body and then by using the right free body. Verify the relationships $\frac{dV}{dx} = -q$ and $\frac{dM}{dx} = V$.

Solution. Using the left free body,

$$q = +2 \text{ klf}$$
$$V = +40.8 - 6 - 2(4 + x) = +26.8 - 2x$$
$$M = +40.8x - 6(4 + x) - \frac{2(4 + x)^2}{2} = -40 + 26.8x - x^2$$

Using the right free body,

$$q = +2 \text{ klf}$$
$$V = -29.2 + 16 + 2(20 - x) = +26.8 - 2x$$
$$M = +29.2(20 - x) - 16(14 - x) - \frac{2(20 - x)^2}{2} = -40 + 26.8x - x^2$$

Note that the same load, shear, and moment equations are obtained and the two differential relationships are satisfied.

Example 5.4. For the beam of Example 5.2 shown again in Fig. 5.10, obtain equations for q, V, and M at a section inside BCD using its distance x from B as the independent variable, first using the left free body and then by using the right free body. Note that an internal hinge is not a section of discontinuity insofar as q, V, and M are concerned, but it is a section of discontinuity on the elastic curve as will be explained later in this chapter.

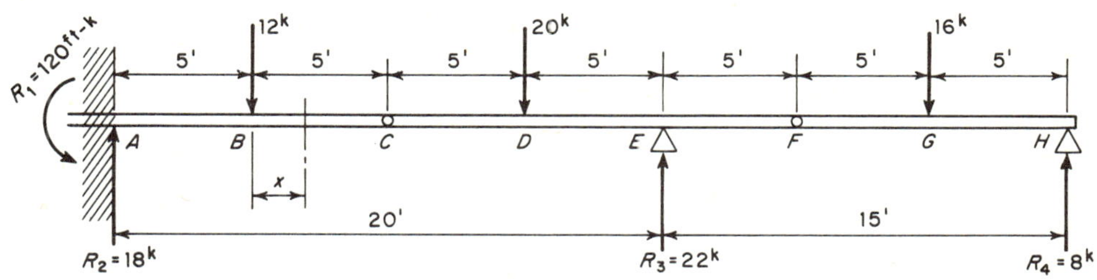

FIG. 5.10. Beam of Example 5.4.

Solution. Using the left free body,

$$q = 0$$
$$V = +18 - 12 = +6 \text{ kips}$$
$$M = -120 + 18(5 + x) - 12x = -30 + 6x$$

Using the right free body,

$$q = 0$$
$$V = -22 - 8 + 20 + 16 = +6 \text{ kips}$$
$$M = 8(30 - x) + 22(15 - x) - 16(25 - x) - 20(10 - x)$$
$$= -30 + 6x$$

EXERCISE 5.4

For the beam of Exercise 5.1 (Fig. 5.5) obtain equations for q, V, and M inside segment CD using its distance x from C as the independent variable.

EXERCISE 5.5

For the beam of Exercise 5.2 (Fig. 5.6) obtain equations for q, V, and M inside segment BC using its distance x from C as the independent variable.

EXERCISE 5.6

For the beam of Exercise 5.3 (Fig. 5.7) obtain equations for q, V, and M inside ABC using its distance x from A as the independent variable.

5.4 Shear and Moment Diagrams

A *shear* or *moment diagram* demonstrates the variation of shear or moment along the length of a beam by showing the plot, usually on a horizontal base line. Shear and moment diagrams are significant in that the designer can visually use them in the design of the strength of the beam.

Since the shear at a section is the sum of the upward forces between the left end and the section, the shear diagram can be simply constructed by starting at the left end and going up and down with the load until the right end is reached. As the slope of the shear curve is the load intensity, the shear variation is horizontally straight over an unloaded segment, and it should slope linearly downward to the right over a uniformly-loaded segment.

The moment diagram may be constructed by first computing its ordinates at the sections of discontinuity and then sketching its curve freehand. From Eq. (5.3.1),

$$\int dM = \int V dx \tag{5.4.1}$$

which means that the increase in moment between one section and an adjacent section to the right is equal to the area of the shear diagram between these two sections. By starting at the left end and successively using Eq. (5.4.1), the moments at as many sections as needed may be computed and plotted. A visual check of the relationship between the moment and shear curves should be made by noting that the slope of the moment curve at any point is equal to the shear intensity at that point.

If the reactions are correctly computed, the shear diagram starts from a zero value at the left end, and terminates with a zero value at the right end; and the moment diagram starts with its known value (if different from zero) at the left end, and terminates with its known value (if different from zero) at the right end. In other words, the shear diagram must close and the moment diagram must close.

Example 5.5. Construct the shear and moment diagrams for the beam of Examples 5.1 or 5.3, shown again in Fig. 5.11a.

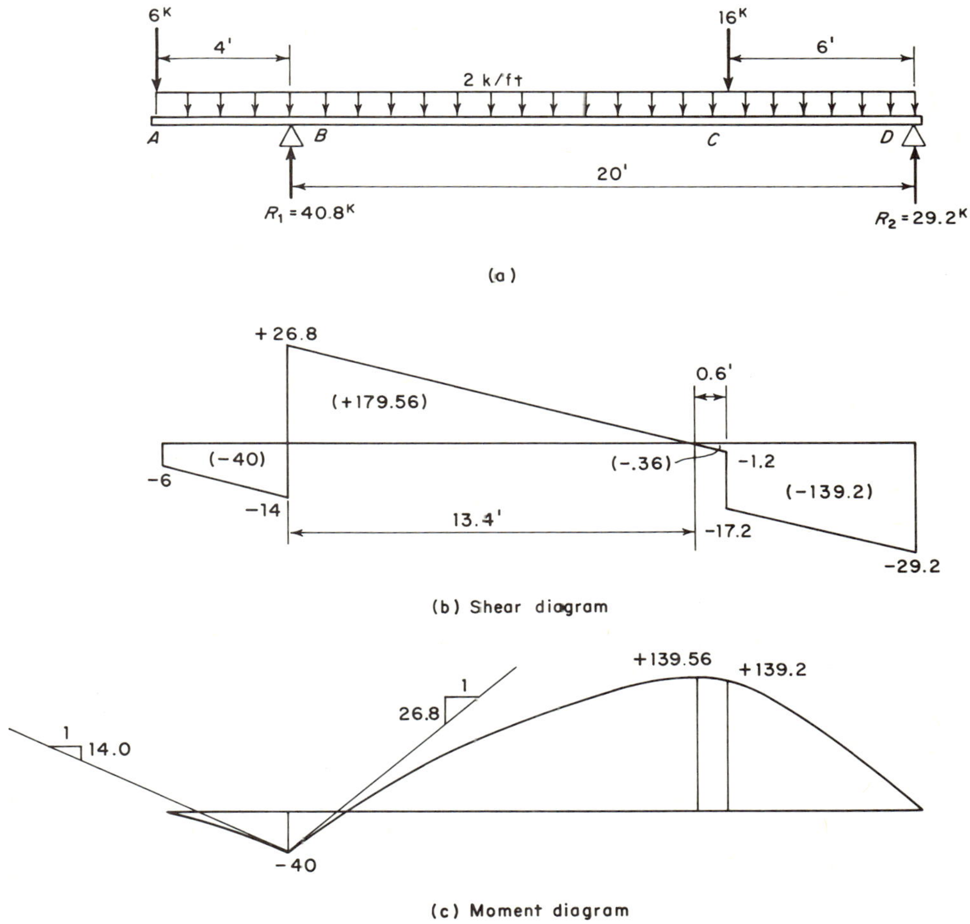

FIG. 5.11. Solution for Example 5.5.

Solution. The shear diagram in Fig. 5.11b is first constructed and found to close up. The point of zero shear is at 26.8/2.0 = 13.4 ft from point *B*, or 1.2/2.0 = 0.6 ft from point *C*. The shear areas are computed to be −40, +179.56, −0.36, and −139.2 ft-kips. Since the change of moments between *A* and *D* is zero, the sum of the four shear areas should be zero.

The tangent to the moment curve is horizontal at the section of zero shear; thus, there is a maximum positive moment at this point. Note in Fig. 5.11c the slope of the moment curve changes from −14.0 to +26.8 at point *B*.

Example 5.6. Construct the shear and moment diagrams for the beam of Examples 5.2 or 5.4, shown again in Fig. 5.12a.

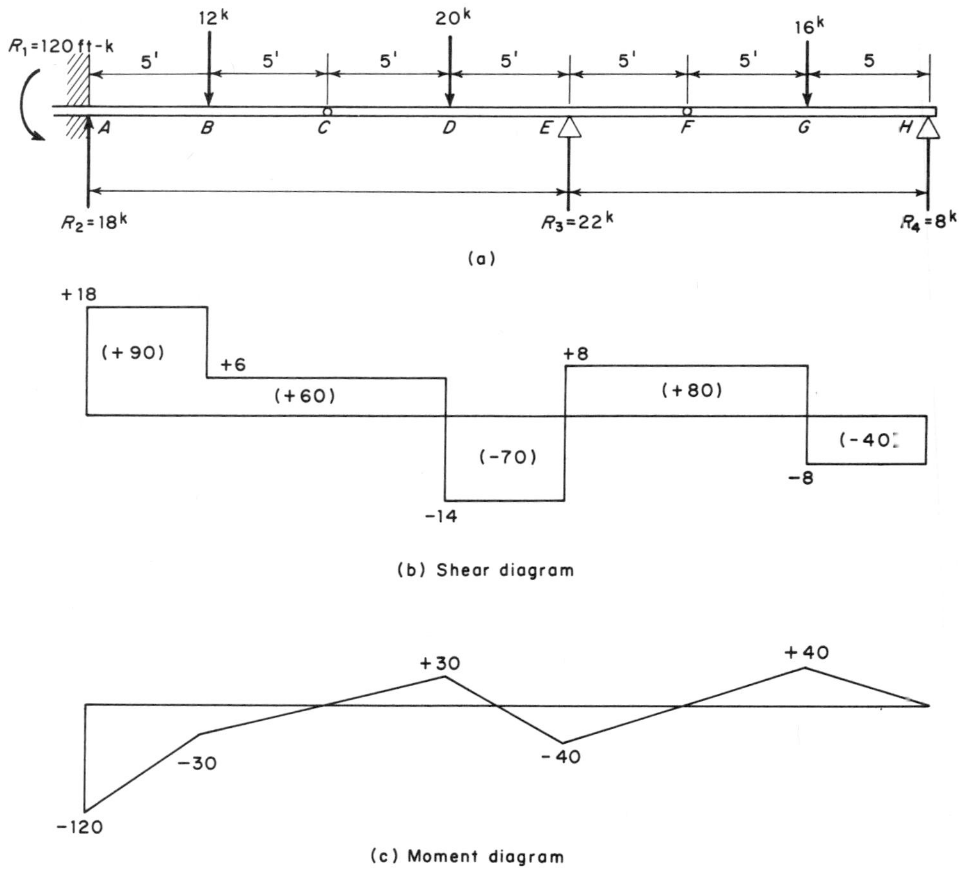

FIG. 5.12. Solution for Example 5.6.

Solution. The shear diagram in Fig. 5.12b is first constructed and found to close up. Since the moment at *A* is −120 ft-kips and the moment at *H* is zero, the change is +120 ft-kips. The sum of the five shear areas in Fig. 5.12b is also +120 ft-kips. So the moment diagram of Fig. 5.12c closes up. Note that the moments are zero at the internal hinges *C* and *F*.

EXERCISE 5.7

Construct the shear and moment diagrams for the beam of Exercises 5.1 or 5.4 (Fig. 5.5).

EXERCISE 5.8

Construct the shear and moment diagrams for the beam of Exercises 5.2 or 5.5 (Fig. 5.6).

EXERCISE 5.9

Construct the shear and moment diagrams for the beam of Exercises 5.3 or 5.6 (Fig. 5.7).

5.5 Elastic Curves of Statically Determinate Beams

The differential equation of the *elastic curve* of a beam is a function of the moment M and the flexural rigidity EI; in symbolic form,

$$\frac{d^2y}{dx^2} = -\frac{M}{EI} \tag{5.5.1}$$

Equation (5.5.1) may be derived by observing Fig. 5.13, in which

$$\frac{d^2y}{dx^2} = \frac{d(dy/dx)}{dx} = -\frac{d\theta}{dx} = -\frac{1}{dx}\left[\frac{\text{elongation of lower fiber}}{c_2}\right]$$
$$= -\frac{1}{dx}\left[\frac{Mc_2}{EI}dx/c_2\right] = -\frac{M}{EI}$$

Note that there is a negative sign before $\frac{d\theta}{dx}$ because with the positive y-axis going in the downward direction, the slope $\frac{dy}{dx}$ is decreasing as x increases.

The equation of the elastic curve can be obtained by integrating Eq. (5.5.1) twice and employing the proper boundary conditions. The elastic curve is discontinuous wherever the moment equation or the boundary condition changes. An internal hinge in a beam does not cause discontinuity in the moment equation, but it does so in the elastic curve equation because the slope may change abruptly at such a hinge.

The double integration method is useful when complete elastic curve equa-

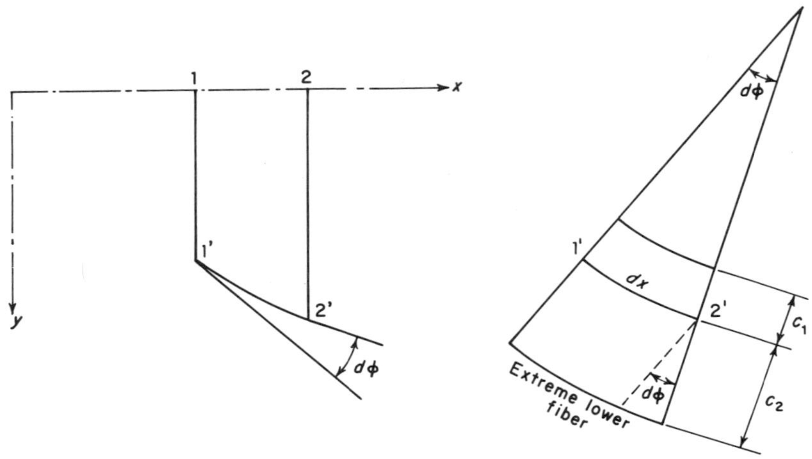

FIG. 5.13. Differential Equation of the Elastic Curve.

tions are required. This method is, however, cumbersome to apply when sections of discontinuity exist because of the large number of integration constants involved. When only the deflections or slopes at some chosen points on the beam are required, the unit-load method or the moment-area method are far more convenient.

5.6 The Moment-Area Method

The *moment-area method* in conjunction with its corollary, the conjugate-beam method, may be conveniently used to determine the deflections or slopes at chosen points on the elastic curve of a beam.

The moment-area method is based on two theorems, known as Moment-Area Theorems I and II:

Moment-Area Theorem I—"The change in slope between any two points on a continuous elastic curve is equal to the $M/(EI)$ area between these two points."

Moment-Area Theorem II—"The distance of a second point on a continuous elastic curve from the tangent at a first point, measured in a direction perpendicular to the original axis of the beam, is equal to the moment of the $M/(EI)$ area between these two points about the second point."

Referring to Fig. 5.14a the two moment-area theorems may be expressed by

(5.6.1) $$\theta_{12} = \int_1^2 \frac{M}{EI} dx$$

(5.6.2) $$\Delta_{12} = \int_1^2 \frac{Mxdx}{EI}$$

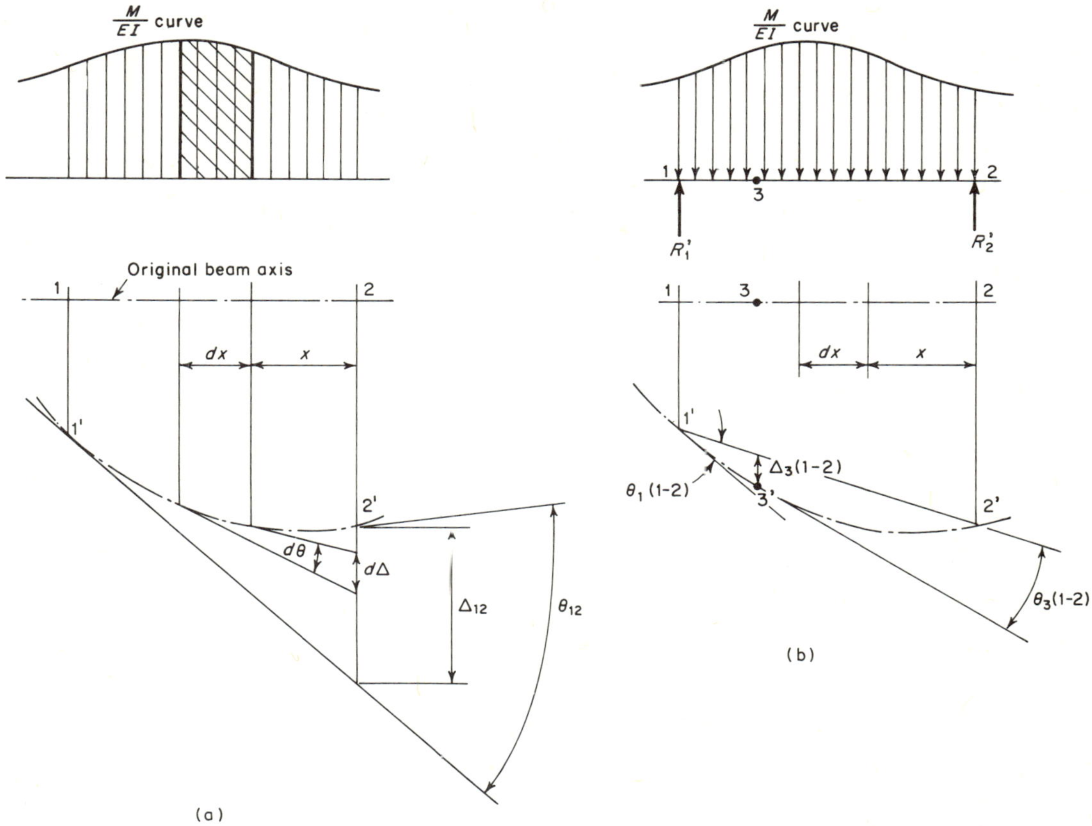

FIG. 5.14. Moment-Area and Conjuagate-Beam Theorems.

The proof of Eq. (5.6.1) may be taken directly from Eq. (5.5.1); or

$$d\theta \text{ in Fig. 5.14a} = \frac{M}{EI}dx$$

from which

$$\theta_{12} = \int_1^2 d\theta = \int_1^2 \frac{M}{EI}dx$$

On basis of the first-order assumption that all deflections are small, the lengths of all slant lines in Fig. 5.14a are equal to their projections on the original beam axis; thus

$$d\Delta = x d\theta$$

from which

$$\Delta_{12} = \int_1^2 d\Delta = \int_1^2 \frac{Mxdx}{EI}$$

The *conjugate-beam method* is based on two corollaries of the two moment-area theorems, commonly known as Conjugate-Beam Theorems I and II:

Conjugate-Beam Theorem I—"The clockwise angle between a straight line joining any two points on a continuous elastic curve and the tangent at a third intermediate point is equal to the shear at the third point in a simple beam equal in span to the distance between the two boundary points and loaded by the $M/(EI)$ area."

Conjugate-Beam Theorem II—"The distance of a third intermediate point from a straight line joining two points on a continuous elastic curve, measured in a direction perpendicular to the original axis of the beam, is equal to the moment at the third point in a simple beam equal in span to the distance between the two boundary points and loaded by the $M/(EI)$ area."

Referring to Fig. 5.14b, the two conjugate-beam theorems may be expressed by

$$\theta_3(1\text{–}2) = V_3' = R_3' - \left[\frac{M}{EI} \text{ area between 1 and 2}\right] \tag{5.6.3}$$

and

$$\begin{aligned} \Delta_3(1\text{–}2) = M_3' &= [R_1' \text{ times distance 1–3}] \\ &\quad - \left[\text{moment of } \frac{M}{EI} \text{ area between 1 and 3 about 3}\right] \end{aligned} \tag{5.6.4}$$

Equations (5.6.3) and (5.6.4) may be proved by means of Eqs. (5.6.1) and (5.6.2). Using the second moment-area theorem,

$$\begin{aligned} \theta_1(1\text{–}2) &= \frac{\text{distance of } 2' \text{ from tangent at } 1'}{\text{distance 1–2}} \\ &= \frac{\text{moment of } M/EI \text{ area between 1 and 2 about 2}}{\text{distance 1–2}} = R_1' \end{aligned}$$

Using the first moment-area theorem,

$$\begin{aligned} \theta_3(1\text{–}2) &= \theta_1(1\text{–}2) - (\text{change in slope between 1 and 3}) \\ &= R_1' - \left(\frac{M}{EI} \text{ area between 1 and 3}\right) = V_3' \end{aligned}$$

which is Eq. (5.6.3). From Fig. 5.14b,

$$\begin{aligned} \Delta_3(1\text{–}2) &= [\theta_1(1\text{–}2) \text{ times distance 1–3}] - [\text{distance of } 3' \text{ from tangent at } 1'] \\ &= [R_1' \text{ times distance 1–3}] - \left[\text{moment of } \frac{M}{EI} \text{ area between 1–3 about 3}\right] \\ &= M_3' \end{aligned}$$

which is Eq. (5.6.4).

The procedure of using the moment-area method to find the deflections and slopes at chosen points on the elastic curve of a beam is as follows:

a) compute the reactions, and draw the shear and moment diagrams;
b) if moment of inertia along the length of the beam is not constant, modify the moment diagram by compressing the ordinate by n times wherever $I = nI_c$, in which I_c is the standard moment of inertia;
c) make a freehand sketch of the elastic curve by observing the sign of the moment over each segment between two adjacent sections of discontinuity;
d) apply the moment-area theorems between two continuous points when both the slope and deflection at one of the two points are known, zero or otherwise; and
e) apply the conjugate-beam theorems between two continuous points when the deflections but not the slopes at both of these points are known, zero or otherwise.

Example 5.7. By the moment-area method compute the slope and deflection at point A of the beam of Examples 5.1, 5.3, or 5.5. Assume constant flexural rigidity EI along the entire length of the beam.

Solution. The reactions and the moment diagram are taken from Examples 5.1 and 5.5 and shown again in Fig. 5.15a and b. Based on the moment diagram, a freehand sketch of the elastic curve is made and shown in Fig. 5.15c. Although the elastic curve between the point A and the point of inflection has to be convex at the top, whether the slope at A is clockwise as shown, and whether the deflection at A is upward as shown, are yet to be confirmed.

The procedure of determining the slope and deflection at A involves: 1) using Conjugate-Beam Theorem I to obtain the slope at B, which is the reaction (or shear) at B of simple beam BD loaded with $M/(EI)$ area; and 2) applying Moment-Area Theorems I and II to segments BA to obtain the slope and deflection at A.

Since EI is constant, the moment diagram may be qualitatively used as the $M/(EI)$ area without modification. Also, the reaction at B to the conjugate simple beam BD may be obtained as the sum of the three parts shown in Fig. 5.15d; thus,

$$\theta_B = \theta_{B1} + \theta_{B2} - \theta_{B3} = R'_{E1} + R'_{B2} - R'_{B3}$$
$$= \frac{291.2}{EI} + \frac{2000}{3EI} - \frac{800}{3EI} = \frac{691.20}{EI} \text{ k-ft}^2 \text{ clockwise}$$

Applying Moment-Area Theorems I and II to segment BA on which the $M/(EI)$ area is shown in two parts in Fig. 5.15e,

$$\theta_A = \theta_B - \left[\frac{M}{EI} \text{ area on } BA\right] = \frac{691.20}{EI} - \left[\frac{48}{EI} + \frac{64}{3EI}\right] = \frac{621.87}{EI} \text{ k-ft}^2 \text{ clockwise}$$

$$\Delta_A = \theta_B(4) - \left[\text{moment of } \frac{M}{EI} \text{ area on } BA \text{ about } A\right]$$
$$= \frac{691.20}{EI}(4) - \left[\frac{48}{EI}\left(\frac{8}{3}\right) + \frac{64}{3EI}(3)\right] = \frac{2572.80}{EI} \text{ k-ft}^3 \text{ upward}$$

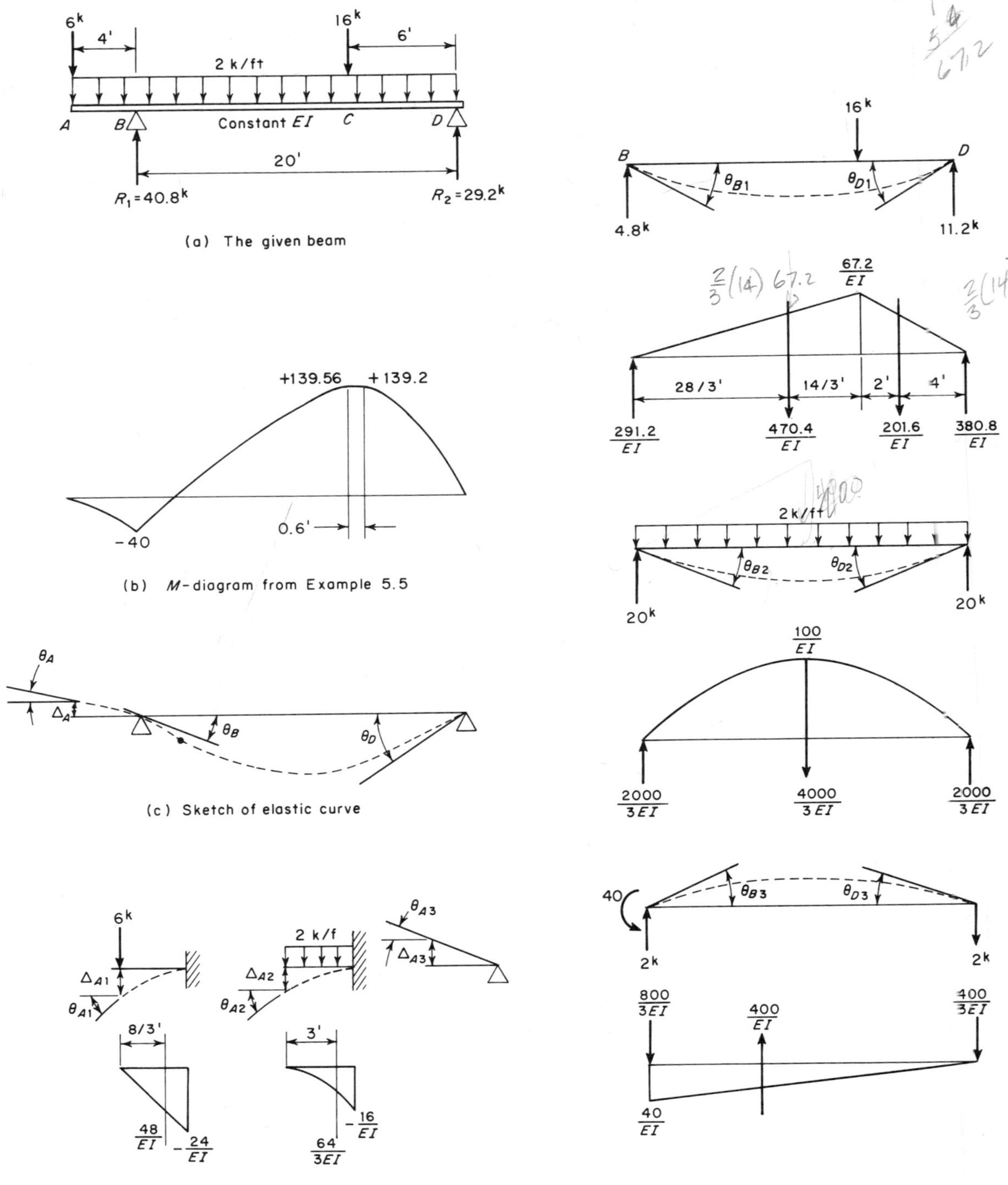

(a) The given beam

(b) M-diagram from Example 5.5

(c) Sketch of elastic curve

(e) Treating AB as a cantilever

(d) Treating BD as a simple beam

FIG. 5.15. Solution for Example 5.7.

Note that θ_B itself is clockwise, and a point travels counterclockwise from B to A on the elastic curve which makes the positive remainder 621.87 k-ft^2/(EI) clockwise. In addition, note that the clockwise rotation of the tangent at point B deflects point A upward, and the point travels downward from this tangent in going from B to A on the elastic curve which makes the positive remainder 2572.80 k-ft^3/(EI) in the upward direction.

Example 5.8. By the moment-area method compute the slope of the left elastic curve at the internal hinge F, the slope of the right elastic curve at the internal hinge F, and the vertical deflection of the internal hinge F in the beam of Examples 5.2, 5.4, and 5.6. Let the moment of inertia of segment $ABCDE$ be $5I_c$ and that of segment EFGH be $2I_c$.

Solution. The reactions and the moment diagram are taken from Examples 5.2 and 5.6 and shown again in Fig. 5.16a and b. Based on the moment diagram a sketch of the elastic curve is done freehand, shown in Fig. 5.16c. It should be pointed out that at this time there is no assurance whether point F' is above or below the original beam axis. All that is known is the degree of concavity or convexity of the curvature along the beam. In other words, corrections of the first assumed sketch of the elastic curve should be continually made as the computations progress. Note, however, the discontinuities in the slope at points C' and F'.

The modified M-diagram of Fig. 5.16d is obtained by compressing the ordinates of the M-diagram in Fig. 5.16b by n, where nI_c is the moment of inertia. In this way, the modified M-diagram would qualitatively represent the $M/(EI)$ area, except the divisor EI_c.

The procedure of determining θ_{FL}, θ_{FR}, and Δ_F as shown in Fig. 5.16c involves: 1) applying Moment-Area Theorem II to segment AC to obtain Δ_c (Fig. 5.16e); 2) applying Conjugate-Beam Theorem I to segment CE to determine the angles between the straight line $C'E$ and the elastic curve at C' and E (Fig. 5.16f); 3) applying both moment-area theorems to segment EF to obtain θ_{FL} and Δ_F (Fig. 5.16g), and 4) applying Conjugate-Beam Theorem I to segment FH to determine the angle between the straight line $F'H$ and the elastic curve at F' (Fig. 5.16h). The detailed computations follow:

For segment AC (Fig. 5.16e),

$$\Delta_c = \frac{1}{EI_c}[\text{moment of areas (1) and (2) about } C]$$

$$= \frac{1}{EI_c}\left[60\left(5 + \frac{10}{3}\right) + 30(5)\right] = \frac{650}{EI_c} \text{ k-ft}^3 \text{ downward}$$

For segment CE (Fig. 5.13f)

$$\phi_C = \phi_{C1} - \phi_{C2} = \frac{1}{EI_c}[\tfrac{1}{2}(30) - \tfrac{1}{6}(20)] = \frac{35}{3EI_c} \text{ k-ft}^2$$

$$\phi_E = \phi_{E2} - \phi_{E1} = \frac{1}{EI_c}[\tfrac{5}{6}(20) - \tfrac{1}{2}(30)] = \frac{5}{3EI_c} \text{ k-ft}^2$$

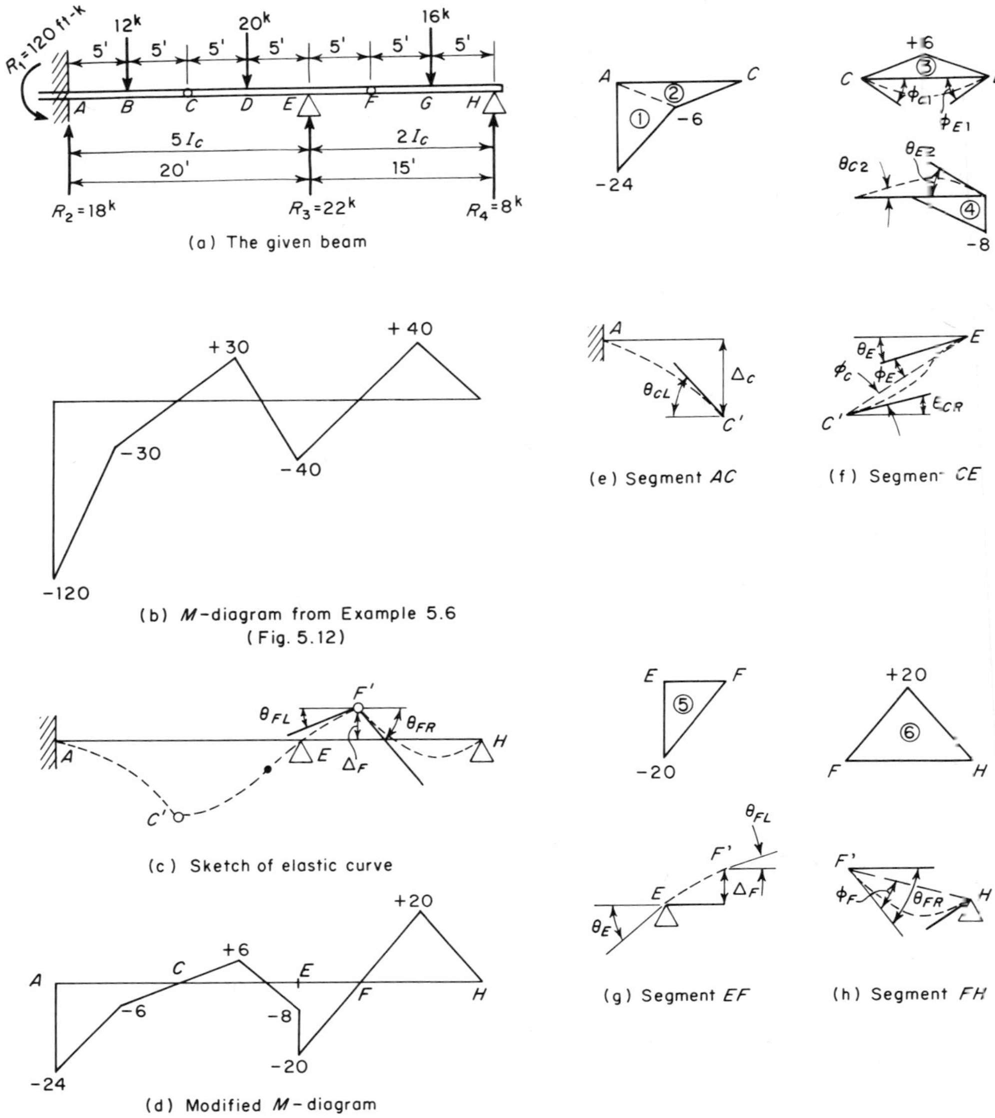

FIG. 5.16. Solution for Example 5.8.

$$\theta_E = [\text{slope of straight line } C'E] - \phi_E = \frac{1}{10}\,\Delta_C - \phi_E$$

$$= \frac{1}{EI_c}\left[65 - \tfrac{5}{3}\right] = \frac{190}{3EI_c} \text{ k-ft}^2 \text{ counterclockwise}$$

For segment EF (Fig. 5.16g),

$$\theta_{FL} = \theta_E - \frac{1}{EI_c}[\text{area (5)}] = \frac{1}{EI_c}\left[\frac{190}{3} - 50\right]$$

$$= \frac{40}{3EI_c} \text{ k-ft}^2 \text{ counterclockwise (answer)}$$

$$\Delta_F = 5\theta_E - \frac{1}{EI_c}[\text{moment of area (5) about } F]$$

$$= \frac{1}{EI_c}\left[5\left(\frac{190}{3}\right) - 50\left(\frac{10}{3}\right)\right] = \frac{150}{EI_c} \text{ k-ft}^3 \text{ upward (answer)}$$

For segment *FH* (Fig. 5.13h),

$$\phi_F = \text{left reaction to conjugate simple beam } FH$$

$$= \frac{1}{EI_c}[\tfrac{1}{2} \text{ of area (6)}] = \frac{50}{EI_c}$$

$$\text{slope of straight line } F'H = \frac{1}{10}\Delta_F = \frac{15}{EI_c}$$

$$\theta_{FR} = [\text{slope of } F'H] + \phi_F$$

$$= \frac{1}{EI_c}[15 + 50] = \frac{65}{EI_c} \text{ k-ft}^2 \text{ clockwise (answer)}$$

Note that the notation θ is used for the angle between the original beam axis and the tangent to the elastic curve; and ϕ is used for the angle between the tangent to the elastic curve and the straight line joining the two boundary points.

EXERCISE 5.10

By the moment-area method compute the slope and deflection at point *F* of the beam of Exercise 5.1 (Fig. 5.5). Assume constant *EI* along the entire length of the beam.

EXERCISE 5.11

By the moment-area method compute the slope and deflection at point *A* of the beam of Exercise 5.2 (Fig. 5.6). Assume constant *EI* along the entire length of the beam.

EXERCISE 5.12

By the moment-area method compute the slopes and the deflection at the internal hinge *B* of the beam of Exercise 5.3 (Fig. 5.7). Assume constant *EI* along the entire length of the beam.

Special note: When a uniform load does not cover the entire length of a simple beam, the moment diagram shown in Fig. 5.17a is the sum of area (1) = *abdca* with straight sides; and area (2) = *cdec*, which is identical with the moment diagram of a simple beam with span equal to *CD*. This may be proved by comparing the moment equation for a point *E* in the beam of Fig. 5.17a

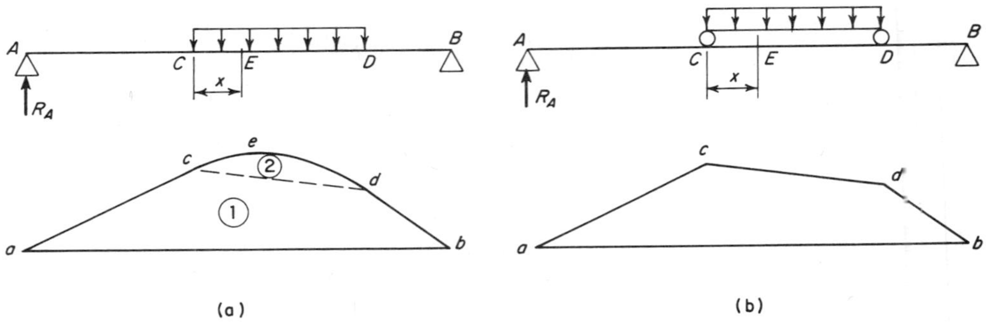

FIG. 5.17. Moment-Area Properties.

with that of Fig. 5.17b; thus

$$M_E(\text{Fig. 5.17a}) = R_A(AC + x) - \frac{wx^2}{2}$$

$$M_E(\text{Fig. 5.17b}) = R_A(AC + x) - \frac{w(CD)}{2}x$$

and

$$M_E(\text{Fig. 5.17a}) - M_E(\text{Fig. 5.17b})$$
$$= \frac{w(CD)}{2}x - \frac{wx^2}{2} = \text{moment in the simple beam } CD$$

5.7 The Unit-Load Method

In Art. 3.6 the *unit-load method* of finding the displacement of a particular joint in a truss in a specific direction has been described. Its derivation is based on the Principle of Virtual Work. In this article the use of the unit-load method to find the deflection and slope at a particular point in a beam will be demonstrated.

Unlike a truss which has a finite number of equivalent springs (or members), a beam may be considered to consist of a compaction of minute springs, each of dx in length in the original state but of $(dx + e)$ in the deformed state. The elongation e in the infinitesimal springs may be due to applied loads, temperature changes, and other physical variables. Thus the compatible state of the beam in Fig. 5.18a has the same meaning as that of the truss in Fig. 3.15a. The contribution of the shear deformation to the geometry of the elastic curve is usually small (except in deep, short beams) and will not be treated in this elementary text.

Now consider the equilibrium state of Fig. 5.18b, or of Fig. 5.18d, in which the unit load (a unit moment is a generalized unit load) of $p_i = 1$ kip or $p_i = 1$ ft-kip is equilibrated by the forces u in the infinitesimal springs. Using a broad interpretation of the words of the Principle of Virtual Work in Art. 3.6, which says "the external work and the internal work, each done by the force in the

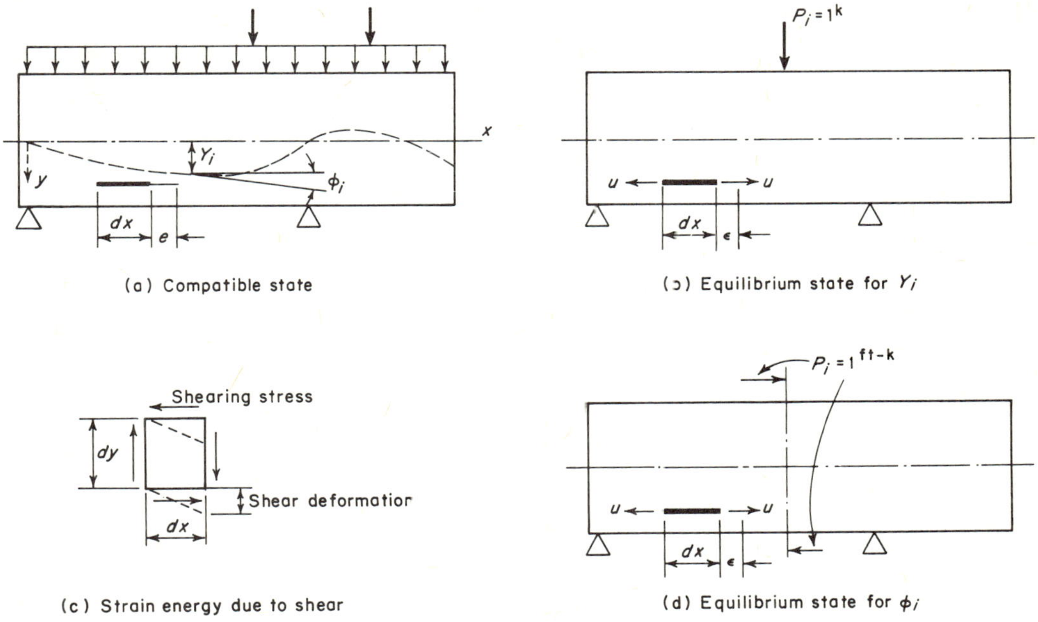

FIG. 5.18. The Unit-Load Method.

'equilibrium state' in going through the deformations of a 'compatible state,' must be equal," one obtains

$$P_i * Y_i = \Sigma(u)(e) \tag{5.7.1}$$

The derivation of Eq. (5.7.1) is more subtle than that of Eq. (3.6.2) because there are no joints to get hold of in a beam. However, one may use the concept of work and energy. Let the deflection in the direction of P_i in the equilibrium state of Fig. 5.18b be y_i, and the elongations in the infinitesimal springs represented by ϵ.

Applying the unit load first to the original beam, neglecting the strain energy due to shear, and equating the external work to the strain energy due to normal flexural stresses,

$$\tfrac{1}{2}P_i * y_i = \Sigma\tfrac{1}{2}(u)(\epsilon) \tag{5.7.2}$$

To this beam already loaded by P_i, the deformations of the compatible state are superimposed. The additional external work done by P_i is P_iY_i and the additional strain energy is $\Sigma(u)(e)$. Equating the total external work and the total strain energy,

$$\tfrac{1}{2}P_i * y_i + P_i * Y_i = \Sigma\tfrac{1}{2}(u)(\epsilon) + \Sigma(u)(e) \tag{5.7.3}$$

Subtracting Eq. (5.7.2) from Eq. (5.7.3),

$$P_i * Y_i = \Sigma(u)(e)$$

which is Eq. (5.7.1). Had the contribution of shear deformation to the geometry of the elastic curve been considered, Eq. (5.7.1) would have included an additional summation term of the strain energy due to shear.

When the elongations in the infinitesimal springs in the compatible state are due to bending moment M,

$$e = \frac{My}{EI}\,dx \tag{5.7.4}$$

Using the symbol m for the bending moment in the equilibrium state,

$$u = \frac{my}{I}\,dA \tag{5.7.5}$$

Substituting Eqs. (5.7.4) and (5.7.5) in Eq. (5.7.1),

$$P_i * Y_i = \Sigma(u)(e) = \iint \frac{My}{EI}\,dx\,\frac{my}{I}\,dA = \int \frac{Mmdx}{EI} \tag{5.7.6}$$

Similarly, when P_i is a unit moment as shown in Fig. 5.18d,

$$P_i * \theta_i = \int \frac{Mmdx}{EI} \tag{5.7.7}$$

Equations (5.7.6) and (5.7.7) are the working formulas for determining the deflection or the slope at one particular point on the elastic curve.

The following two examples will show that exactly identical results are obtained by the unit-load and moment-area methods. In general, a problem should be first solved by the moment-area method, and a check is then made by the unit-load method.

Example 5.9. Solve Example 5.7 by the unit-load method.

Solution. The reactions to the given beam should be first computed but in this case they are taken from the previous example and shown again in Fig. 5.19a. The unit moment and the unit load are applied at point A and the equilibrium diagrams are shown in Fig. 5.19b and c, respectively.

The moment equations for each of the three segments between discontinuous sections, and for each of the three equilibrium states, are

1) Segment AB, origin at A, limits from 0 to 4:

$$M = -6x - x^2$$
$$m \text{ for } \theta_A = -1.0$$
$$m \text{ for } \Delta_A = -x$$

2) Segment BC, origin at D, limits from 6 to 20,

$$M = 29.2x - x^2 - 16(x - 6)$$
$$m \text{ for } \theta_A = -0.05x$$
$$m \text{ for } \Delta_A = -0.20x$$

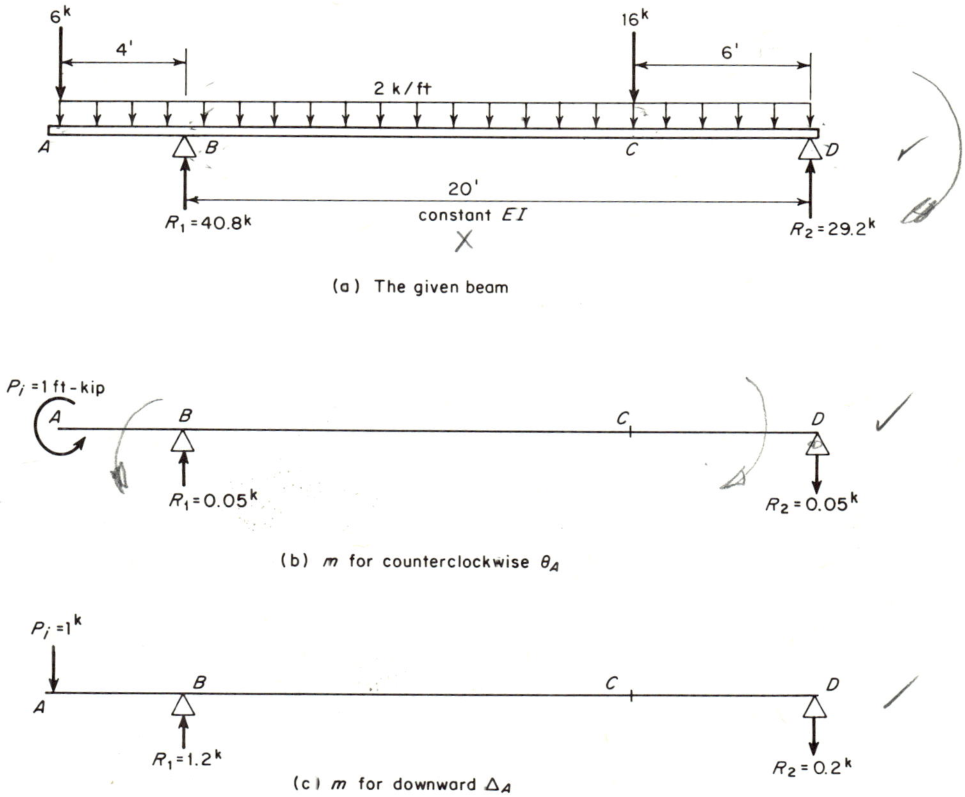

FIG. 5.19. Solution for Example 5.9.

3) Segment CD, origin at D, limits from 0 to 6,

$$M = 29.2x - x^2$$
$$m \text{ for } \theta_A = -0.05x$$
$$m \text{ for } \Delta_A = -0.20x$$

Applying the working formulas (5.7.6) and (5.7.7) in the unit-load method,

$$\begin{aligned} EI\theta_A &= \int (M)(m \text{ for } \theta_A)\, dx \\ &= \int_0^4 (-6x - x^2)(-1.0)\, dx + \int_6^{20} [29.2x - x^2 - 16(x - 6)](-0.05x)\, dx \\ &\quad + \int_0^6 (29.2x - x^2)(-0.05x)\, dx \\ &= (+69.33) + (-88.92) + (-602.28) = -621.87 \end{aligned}$$

$$\theta_A = \frac{621.87}{EI} \text{ k-ft}^2 \text{ clockwise}$$

$$EI\Delta_A = \int (M)(m \text{ for } \Delta_A)\,dx$$

$$= \int_0^4 (-6x - x^2)(-x)\,dx + \int_6^{20} [29.2 - x^2 - 16(x-6)](-0.20x)\,dx$$

$$+ \int_0^6 (29.2x - x^2)(-0.20x)\,dx$$

$$= (+192) + (-355.68) + (-2409.12) = -2572.8$$

$$\Delta_A = \frac{2572.8}{EI} \text{ k-ft}^3 \text{ upward}$$

Example 5.10. Solve Example 5.8 by the unit-load method.

Solution. The reactions R_1 to R_4 in the given beam should be first computed, but in this case they are taken from the previous example and shown again in Fig. 5.20a. Using the notations θ_{FL} and θ_{FR}, respectively, for the slopes at F of the elastic curves to the left and right of F, the unit moments are applied as shown in Fig. 5.20b and c. Likewise, the unit load is applied at F in Fig. 5.20d for finding the deflection Δ_F at F.

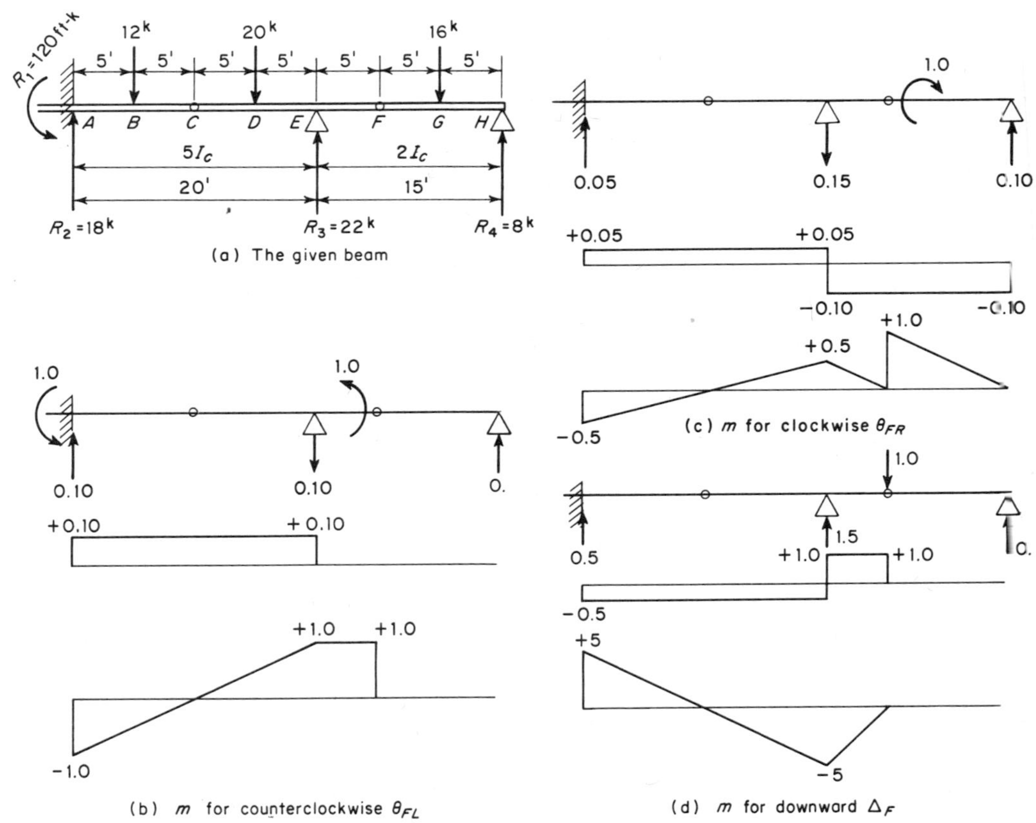

FIG. 5.20. Solution for Example 5.10.

In each case of Fig. 5.20b, c, and d, the reactions are computed by successively using segments FGH, $CDEF$, and ABC as free bodies. The shear and moment diagrams are shown as the means for checking. The moment equations for each of the six segments between discontinuous sections are

Segment	n	Origin	Limits	M	m for θ_{FL}	m for θ_{FR}	m for Δ_F
AB	5	A	0 to 5	$-120 + 18x$	$-1 + 0.10x$	$-0.5 + 0.05x$	$+5 - 0.50x$
BCD	5	A	5 to 15	$-120 + 18x$ $-12(x - 5)$	$-1 + 0.10x$	$-0.5 + 0.05x$	$+5 - 0.50x$
DE	5	A	15 to 20	$-120 + 18x$ $-12(x - 5)$ $-20(x - 15)$	$-1 - 0.10x$	$-0.5 + 0.05x$	$+5 - 0.50x$
EF	2	H	10 to 15	$8x - 16(x - 5)$	$+1$	$-1 + 0.10x$	$-(x - 10)$
FG	2	H	5 to 10	$8x - 16(x - 5)$	0	$+0.10x$	0
GH	2	H	0 to 5	$8x$	0	$+0.10x$	0

Applying the working formulas (5.7.6) and (5.7.7) in the unit-load method,

$$\begin{aligned}
EI_c\theta_{FL} &= \tfrac{1}{5}\int_0^5 (-120 + 18x)(-1 + 0.10x)\,dx \\
&\quad + \tfrac{1}{5}\int_5^{15} [-120 + 18x - 12(x - 5)](-1 + 0.10x)\,dx \\
&\quad + \tfrac{1}{5}\int_{15}^{20} [-120 + 18x - 12(x - 5) - 20(x - 15)](-1 + 0.10x)\,dx \\
&\quad + \tfrac{1}{2}\int_{10}^{15} [8x - 16(x - 5)](-1)\,dx \\
&= (+60) + (+10) + (-\tfrac{20}{3}) + (-50) = +\tfrac{40}{3}
\end{aligned}$$

$$\theta_{FL} = \frac{40}{3EI_c}\ \text{k-ft}^2 \text{ counterclockwise}$$

$$\begin{aligned}
EI_c\theta_{FR} &= \tfrac{1}{5}\int_0^5 (-120 + 18x)(-0.5 + 0.05x)\,dx \\
&\quad + \tfrac{1}{5}\int_5^{15} [-120 + 18x - 12(x - 5)](-0.5 + 0.05x)\,dx \\
&\quad + \tfrac{1}{5}\int_{15}^{20} [-120 + 18x - 12(x - 5) - 20(x - 15)](-0.5 + 0.05x)\,dx \\
&\quad + \tfrac{1}{2}\int_{10}^{15} [8x - 16(x - 5)](-1 + 0.10x)\,dx \\
&\quad + \tfrac{1}{2}\int_{5}^{10} [8x - 16(x - 5)](+0.10x)\,dx \\
&\quad + \tfrac{1}{2}\int_{0}^{5} (8x)(+0.10x)\,dx
\end{aligned}$$

$$= (+30) + (+5) + (-\tfrac{10}{3}) + (-\tfrac{50}{3}) + (+\tfrac{100}{3}) + (+\tfrac{50}{3}) = +65$$

$$\theta_{FR} = \frac{65}{EI_c} \text{ k-ft}^2 \text{ clockwise}$$

$$EI_c\Delta_F = \tfrac{1}{5}\int_0^5 (-120 + 18x)(+5 - 0.50x)\,dx$$
$$+ \tfrac{1}{5}\int_5^{15} [-120 + 18x - 12(x-5)](+5 - 0.50x)\,dx$$
$$+ \tfrac{1}{5}\int_{15}^{20} [-120 + 18x - 12(x-5) - 20(x-15)](+5 - 0.50x)\,dx$$
$$+ \tfrac{1}{2}\int_{10}^{15} [8x - 16(x-5)][-(x-10)]\,dx$$
$$= (-300) + (-50) + (+\tfrac{100}{3}) + (+\tfrac{500}{3}) = -150$$

$$\Delta_F = \frac{150}{EI_c} \text{ k-ft}^3 \text{ upward}$$

EXERCISE 5.13

Solve Exercise 5.10 by the unit-load method.

EXERCISE 5.14

Solve Exercise 5.11 by the unit-load method.

EXERCISE 5.15

Solve Exercise 5.12 by the unit-load method.

5.8 Concluding Remarks

The emphasis of this chapter has been on the quantitative evaluation of shear and moment diagrams as well as on properties of the elastic curve of statically determinate beams. It will be shown in subsequent chapters that this material is basic in obtaining information at points within each span of a statically indeterminate continuous beam or between ends of each member in a statically indeterminate rigid frame. The reason for this is that whether the force method or the displacement method is used, only forces or moments at the ends of each span or member are determined in the analysis of the statically indeterminate structure as a whole.

The moment-area and unit-load methods have been treated in general and two numerical examples given. Students are urged to practice further with these two methods by solving for deflections and slopes in typical regular beams such as simple beams loaded uniformly or by a concentrated load at midspan; or cantilever beams loaded uniformly or by a concentrated load at the free ends.

Statically Indeterminate Beams

6

6.1 Degree of Indeterminacy

The degree of indeterminacy of a statically indeterminate beam is equal to the number of external reactions minus two when no internal hinges are present. The above can be stated in accordance with the fact that statics offers only two independent equations for the equilibrium of a system of coplanar parallel forces. Occasionally, actual internal hinges may be inserted in continuous beams for reasons of design (although more often such a case is encountered in some intermediate stage of limit analysis); then, the degree of indeterminacy is reduced by one for each internal hinge. Thus, a general expression for the degree of indeterminacy of a statically indeterminate beam is

(6.1.1) $$i = NR - NIH - 2$$

in which NR and NIH are the numbers of reactions and internal hinges, respectively. As a typical example, the degree of indeterminacy of the beam in Fig. 6.1a is three, but that in Fig. 6.1b is two.

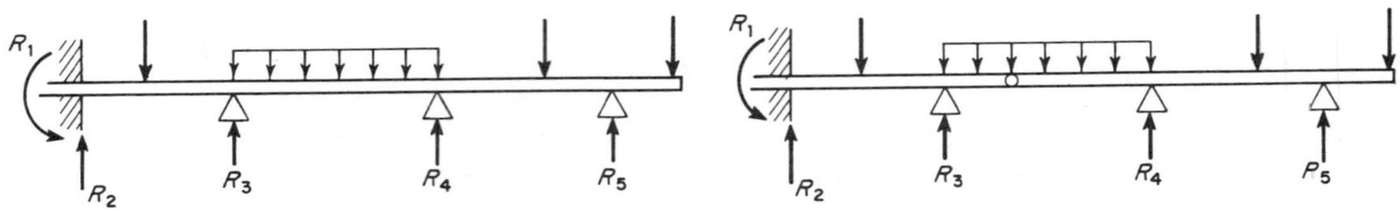

(a) $i = 3$ (b) $i = 2$

FIG. 6.1. Statically Indeterminate Beams.

6.2 Degree of Freedom

A beam, whether statically determinate or indeterminate, may be separated into a finite number of members, at points which are called *joints*. For the purpose of analysis, any point at which there is either an unknown reaction or an internal hinge must be taken as a joint. Therefore, the minimum number of joints in the beam of Fig. 6.1a is four, but that of Fig. 6.1b is five. Additional joints may be regarded as lying anywhere along the beam, each joint increasing the number of members by one.

A joint in a beam may be either fixed against rotation, as at a built-in end, or free to rotate under load. Further, it may be either prevented from transverse displacement, as at a support, or free to deflect under load. Thus, the usual degree of freedom at a joint in a beam is two, except at an internal hinge at which there are two unknown rotations instead of the usual one.

The degree of freedom for a whole beam is the total number of unknown rotations and transverse displacements at all of its joints. Since superfluous joints may be added in any analysis, the degree of freedom NP is a function of the number of joints NJ. Thus, for the beam in Fig. 6.2, if point E is not considered as a joint, as in Fig. 6.2a, the degree of freedom NP is three; but if point E is taken as a joint, as in Fig. 6.2b, NP becomes five.

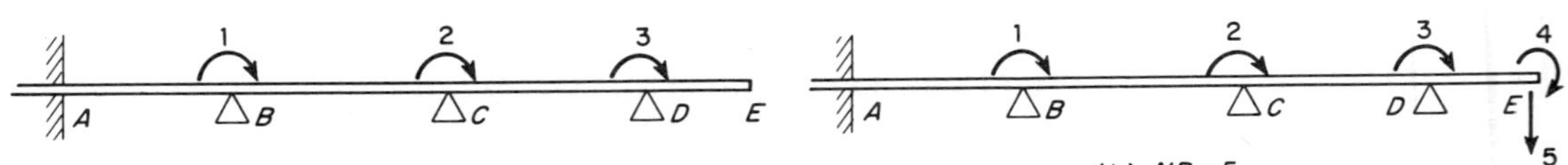

(a) $NP = 3$ (b) $NP = 5$

FIG. 6.2. Degree of Freedom of a Beam.

It will be shown later that when computers are used in the analysis, the selection of joint locations and the resulting degree of freedom depends on the information required in the computer output.

6.3 Force and Displacement Methods

The concept of force and displacement methods, previously discussed in Chapter 4 in relation to trusses, applies equally well to beam analysis.

The force method can be related only to the analysis of statically indeterminate beams where redundant forces, equal in number to the degree of indeterminacy, are taken as the primary unknowns and, as such, are determined first by use of the compatibility conditions in the directions of the redundants. The chosen redundants may consist of external reactions only, of bending moments at the supports only, or of a combination of the two. The first two methods are the ones usually used; in fact, the second method forms the basis of the well-known *three-moment equation.*

The displacement method may be used to analyze both statically determinate and indeterminate beams. In this method, the joint displacements, equal in number to the degree of freedom, are taken as the primary unknowns and, as such, are determined first by use of the equilibrium conditions in the directions of the unknown joint displacements. When conventional longhand algebraic equations are used, this method is known as the *slope-deflection method.* However, when matrix notations are used in the formulation, this method has more recently come to be known as the *matrix-displacement method.* The latter is particularly significant because the matrix operations can be conveniently processed by the digital computer. The well-known *moment-distribution method* developed by Professor Hardy Cross* involves essentially the solution of the simultaneous equations generated in the slope-deflection method by a relaxation procedure.

The remainder of this chapter will present in detail the two force methods and the three displacement methods that have been introduced here.

6.4 Force Method—Reactions as Redundants

When the chosen redundants in the force method of analysis are reactions, the physical restraints associated with the redundant reactions should be removed and the original beam replaced by a "degenerate" statically determinate beam acted upon by the applied loads and the redundant reacting forces. For instance, the fixed-ended beam of Fig. 6.3a or d may be regarded as a degenerate simple beam of Fig. 6.3b or e, or as a degenerate cantilever beam of Fig. 6.3c or f. Since a simple beam is structurally more sensible than a cantilever beam, the first choice should yield better conditioned simultaneous equations in a numerical solution.

*Hardy Cross: "Analysis of Continuous Frames by Distributing Fixed End Moments," *Transactions*, ASCE, 1932.

The redundant reacting forces are determined first from the fact that the rotational or transverse displacements in the directions of the redundants should be zero. Thus, in the case of Fig. 6.3b or e, R_1 and R_2 should be such that they, together with the applied loads, cause zero slopes at both ends of the simple beam; but in the case of Fig. 6.3c or f, R_2 and R_4 should be such that they, together with the applied loads, cause zero slope and zero deflection at the free end of the cantilever.

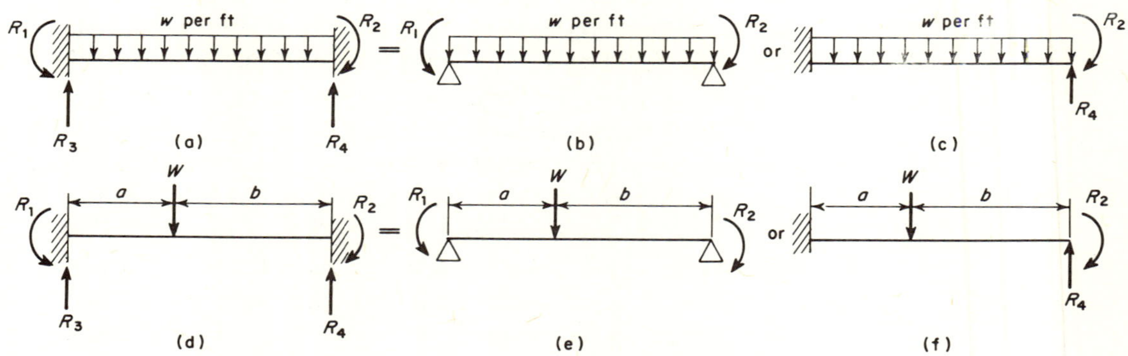

FIG. 6.3. Reactions as Redundants.

After the redundant reacting forces have been obtained, the remaining reactions are determined from the conditions of statics and the shear and moment diagrams of the original beam may then be found.

Generally, a solution can be first obtained by using one convenient set of redundants, and a check is then made by confirming that the compatibility conditions are satisfied in a different degenerate beam. In the case of the fixed-ended beam, R_1 and R_2 in Fig. 6.3a or d may first be obtained by using the degenerate simple beam, and a check is made by verifying that R_2 and R_4 will cause zero slope and zero deflection at the free end of the degenerate cantilever beam.

Example 6.1. Solve for the reactions to a fixed-ended beam subjected to uniform load by using a degenerate simple beam, and check the solution by using a degenerate cantilever beam.

Solution

(a) *Obtain solution from degenerate simple beam (Fig. 6.4a).* Due to symmetry, it is known in advance that R_1 and R_2 are equal. By the conjugate-beam method,

$$\theta_A = \theta_{A1} + \theta_{A2} = \frac{wL^3}{24EI} - \frac{(R_1 \text{ or } R_2)L}{2EI} = 0$$

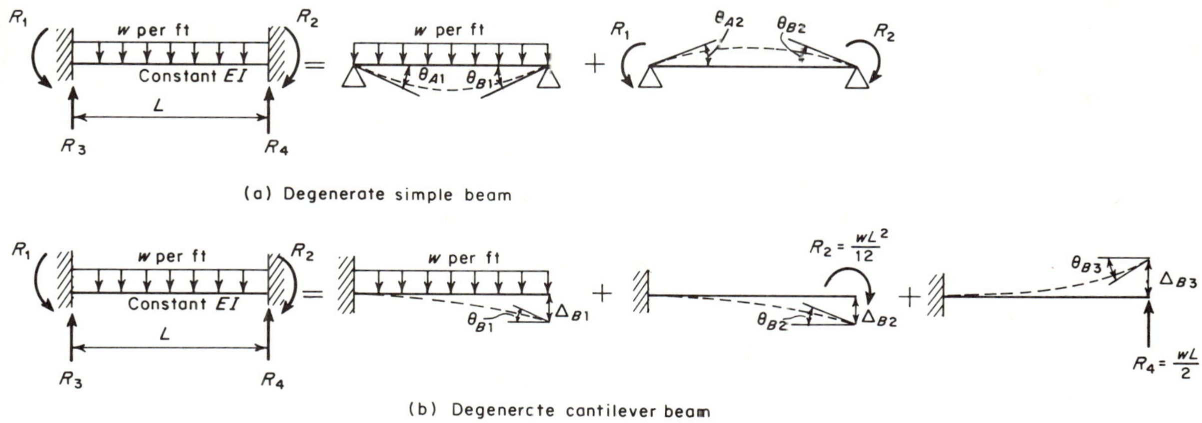

FIG. 6.4. A Fixed-Ended Beam Subjected to a Uniform Load.

from which

$$R_1 \text{ or } R_2 = \frac{wL^2}{12}$$

By symmetry,

$$R_3 \text{ and } R_4 = \frac{wL}{2}$$

(b) *Check solution by degenerate cantilever beam* (*Fig. 6.4b*). By the moment-area method,

$$\theta_B = \theta_{B1} + \theta_{B2} - \theta_{B3} = \frac{wL^3}{6EI} + \frac{wL^3}{12EI} - \frac{wL^3}{4EI} = 0 \qquad \text{(check)}$$

$$\Delta_B = \Delta_{B1} + \Delta_{B2} - \Delta_{B3} = \frac{wL^4}{8EI} + \frac{wL^4}{24EI} - \frac{wL^4}{6EI} = 0 \qquad \text{(check)}$$

Example 6.2. Solve for the reactions to a fixed-ended beam subjected to a concentrated load by using a degenerate simple beam and check the solution by using a degenerate cantilever beam.

Solution

(a) *Obtain solution from degenerate simple beam* (*Fig. 6.5a*). By the conjugate-beam method,

$$\theta_A = \theta_{A1} - \theta_{A2} - \theta_{A3} = \frac{Wab}{2EI}\left(\frac{L+b}{3L}\right) - \frac{R_1L}{2EI}\left(\frac{2}{3}\right) - \frac{R_2L}{2EI}\left(\frac{1}{3}\right) = 0$$

$$\theta_B = \theta_{B1} - \theta_{B2} - \theta_{B3} = \frac{Wab}{2EI}\left(\frac{L+a}{3L}\right) - \frac{R_1L}{2EI}\left(\frac{1}{3}\right) - \frac{R_2L}{2EI}\left(\frac{2}{3}\right) = 0$$

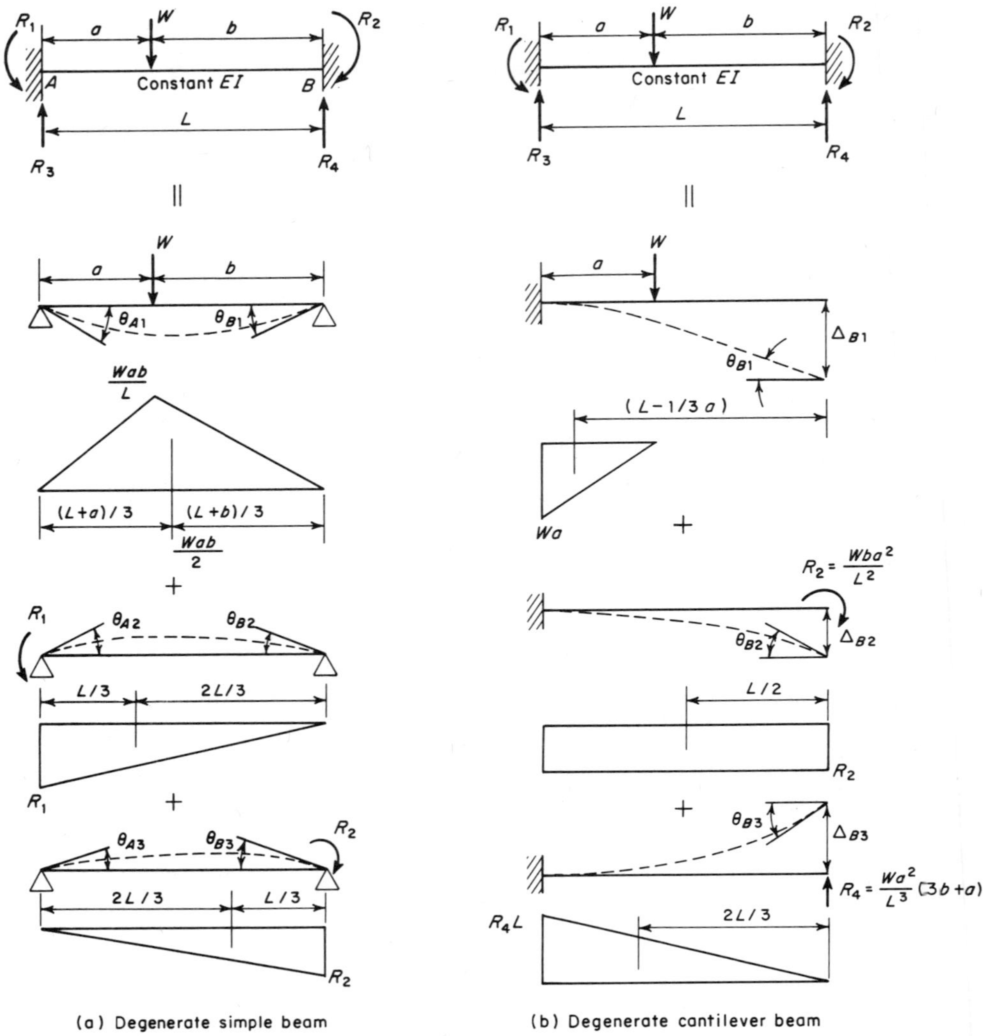

FIG. 6.5. A Fixed-Ended Beam Subjected to a Concentrated Load.

Solving the above two simultaneous equations,

$$R_1 = \frac{Wab^2}{L^2} \qquad R_2 = \frac{Wba^2}{L^2}$$

From the two equations of statics,

$$R_3 = \frac{Wb}{L} + \frac{R_1 - R_2}{L} = \frac{Wb^2}{L^3}(3a + b)$$

$$R_4 = \frac{Wa}{L} - \frac{R_1 - R_2}{L} = \frac{Wa^2}{L^3}(3b + a)$$

(b) *Check solution by degenerate cantilever beam (Fig. 6.5b).* By the moment-area method,

$$\theta_B = \theta_{B1} + \theta_{B2} - \theta_{B3} = \frac{Wa^2}{2EI} + \frac{R_2 L}{EI} - \frac{R_4 L^2}{2EI} = \frac{Wa^2}{2EI} + \frac{Wba^2}{L\,EI} = \frac{Wa^3(3b+a)}{2L\,EI}$$

$$= \frac{W}{2L\,EI}(La^2 + 2a^2b - 3a^2b - a^3) = 0 \quad \text{(check)}$$

$$\Delta_B = \Delta_{B1} + \Delta_{B2} - \Delta_{B3} = \frac{Wa^2}{2EI}\left(L - \frac{1}{3}a\right) + \frac{Wba^2}{L\,EI}\left(\frac{L}{2}\right) - \frac{Wa^2(3b+a)}{2L\,EI}\left(\frac{2}{3}L\right)$$

$$= \frac{W}{6EI}(3La^2 - a^3 + 3a^2b - 6a^2b - 2a^3) = 0 \quad \text{(check)}$$

EXERCISES 6.1 and 6.2

Solve for the reactions to the beams as shown in Figs. 6.6 and 6.7 by using a degenerate simple beam and check the solution by using a degenerate cantilever beam.

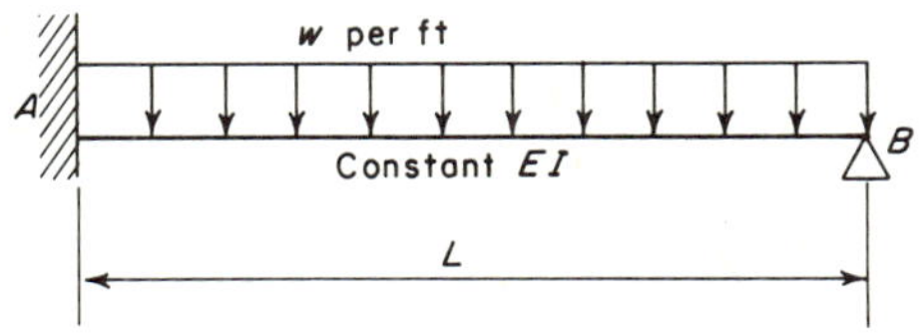

FIG. 6.6. Exercise 6.1.

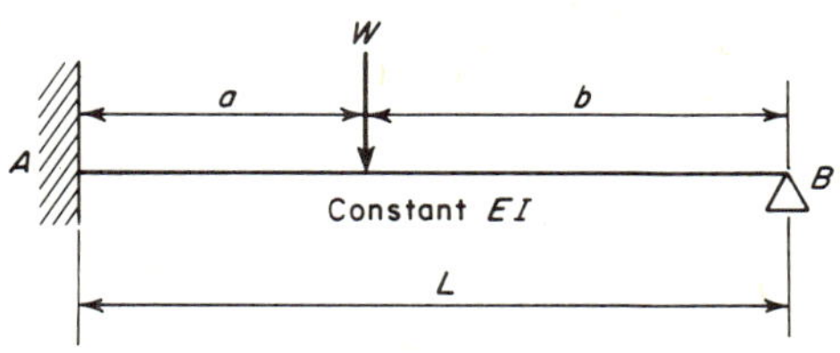

FIG. 6.7. Exercise 6.2.

EXERCISES 6.3 and 6.4

Solve for the reactions to the beams as shown in Figs. 6.8 and 6.9 by using a degenerate simple beam supported at A and C, and check the solution by using a degenerate overhanging beam supported at A and B.

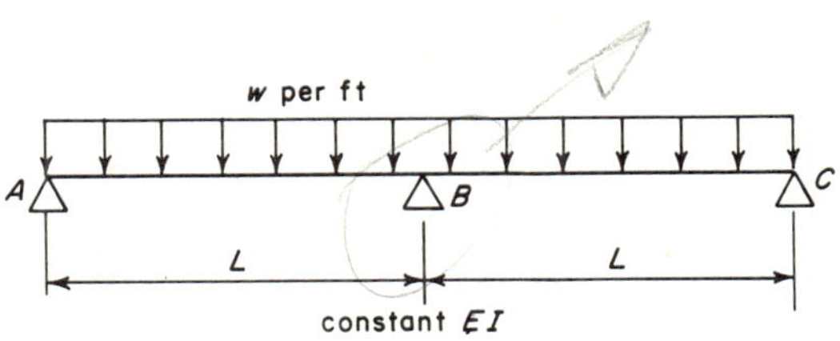

FIG. 6.8. Exercise 6.3.

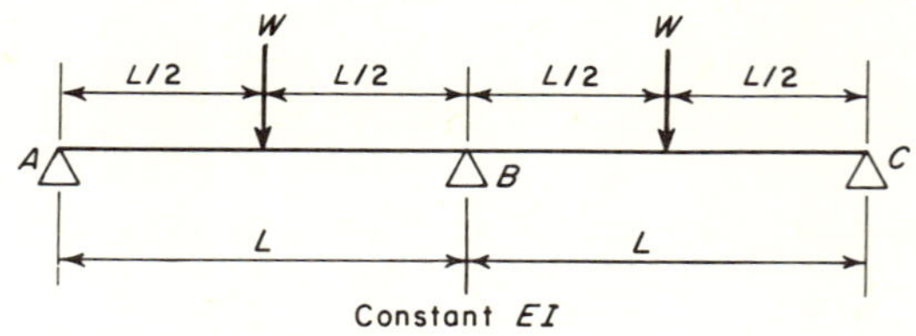

FIG. 6.9. Exercise 6.4.

6.5 Force Method—The Three-Moment Equation

When the bending moments at the supports of a continuous beam are used as the redundants, the compatibility conditions are those of continuity at the supports: *The slope of the elastic curve at the right end of the left span should be equal to the slope at the left end of the right span.*

Consider the two adjacent spans AB and BC in Fig. 6.10. Each span, when taken as a separate free body, is subjected to the load applied on the span and the bending moments at its ends. The moment diagram due to the load on the

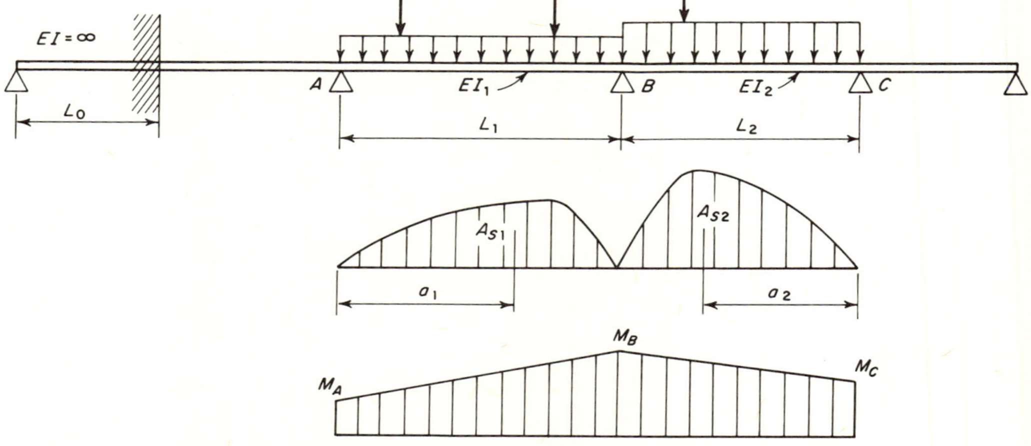

FIG. 6.10. The Three-Moment Equation.

span is the same as if the load is applied to a simple beam, such as A_{S1} and A_{S2} in Fig. 6.10. The distances a_1 and a_2 are measured from the centroids of the simple-beam bending moment diagrams to the supports A and C, respectively. The bending moments at the ends of each span are the unknown redundants; and as such, they are assumed to be positive, causing the variation from M_A to M_B in span AB and from M_B to M_C in span BC. This is demonstrated in Fig. 6.10.

The counterclockwise slope at the right end of span AB is the reaction to the conjugate simple beam AB; or

$$[\theta \text{ at point } B \text{ of span } AB] = \frac{A_{S1}}{EI_1}\frac{a_1}{L_1} + \frac{M_A L_1}{6EI_1} + \frac{M_B L_1}{3EI_1} \tag{6.5.1}$$

Likewise, the clockwise slope at the left end of span BC is

$$[\theta \text{ at point } B \text{ of span } BC] = \frac{A_{S2}}{EI_2}\frac{a_2}{L_2} + \frac{M_B L_2}{3EI_2} + \frac{M_C L_2}{6EI_2} \tag{6.5.2}$$

For continuity in slope at the intermediate support B, Eq. (6.5.1) should be equal to the negative of Eq. (6.5.2); thus

$$M_A\left(\frac{L_1}{6EI_1}\right) + M_B\left(\frac{L_1}{3EI_1} + \frac{L_2}{3EI_2}\right) + M_C\left(\frac{L_2}{6EI_2}\right) = -\frac{A_{S1}}{EI_1}\frac{a_1}{L_1} - \frac{A_{S2}}{EI_2}\frac{a_2}{L_2} \tag{6.5.3}$$

Equation (6.5.3) is called the *three-moment equation* because it is the compatibility condition connecting the three bending moments at three adjacent supports in a continuous beam.

Whenever there is a fixed exterior support, a fictitious span of any arbitrary length L_0, but with infinite flexural rigidity ($EI = \infty$), may be added beyond the fixed support, as shown in Fig. 6.10. In this way, there are always as many three-moment equations as unknown bending moments at the supports.

Example 6.3. Analyze the continuous beam shown in Fig. 6.11 by using the three-moment equation.

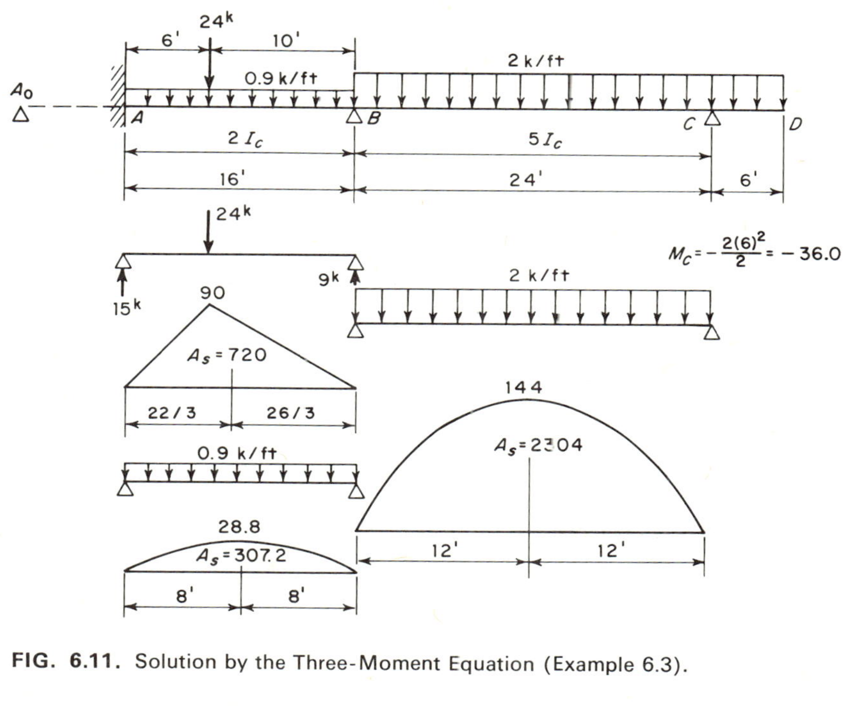

FIG. 6.11. Solution by the Three-Moment Equation (Example 6.3).

Solution. The given beam is statically indeterminate to the second degree. When the bending moments at A and B are used as the redundants, the compatibility conditions are those of continuity of the elastic curve at A and B. Upon adding the fictitious span beyond the fixed support A, two "three-moment" equations (6.5.3) may be written, first for spans A_0A and AB, and then for spans AB and BC. The bending moment at C is negative and its value is -36.0 ft-kips.

The three-moment equation for spans A_0A and AB is

$$M_{A0}\left[\frac{L_0}{6(\infty)}\right] + M_A\left[\frac{L_0}{3(\infty)} + \frac{16}{3E(2I_C)}\right] + M_B\left[\frac{16}{6E(2I_C)}\right] = -\frac{720(26/3)}{E(2I_C)(16)} - \frac{307.2(8)}{E(2I_C)(16)}$$

and that for spans AB and BC is

$$M_A\left[\frac{16}{6E(2I_C)}\right] + M_B\left[\frac{16}{3E(2I_C)} + \frac{24}{3E(5I_C)}\right] + M_C\left[\frac{24}{6E(5I_C)}\right] = -\frac{720(22/3)}{E(2I_C)(16)} - \frac{307.2(8)}{E(2I_C)(16)} - \frac{2304(12)}{E(5I_C)(24)}$$

Noting that anything divided by infinity is zero and cancelling EI_C from the denominator in each term, the above equations become

$$M_A[\tfrac{8}{3}] + M_B[\tfrac{4}{3}] = -271.8$$
$$M_A[\tfrac{4}{3}] + M_B[\tfrac{64}{15}] = -443.4$$

Solving,

$$M_A = -59.22 \text{ ft-kips}$$
$$M_B = -85.42 \text{ ft-kips}$$

The shear and moment diagrams for spans AB and BC, as well as for the overhang CD, are shown in Fig. 6.12a. For span AB, the negative bending moments of 59.22 and 85.42 ft-kips are first placed at the ends. The end reactions of $+7.20$ and $+7.20$ kips are due to the uniform load on AB; $+15.00$ and $+9.00$ kips, due to the concentrated load on AB; and -1.638 and $+1.638$ kips, due to the unbalanced end moment of $85.42 - 59.22 = 26.20$ ft-kips clockwise, which requires a counterclockwise reacting couple made of an upward force at B and a downward force at A.

Two continuity checks may be made by applying the conjugate-beam method to spans AB and BC, as shown in Fig. 6.12b. The two checks are

1) θ_A (span AB) $= 0$,

$$\theta_A = \frac{1}{EI_C}(+195 + 76.8 - 157.92 - 113.89) = \frac{1}{EI_C}(271.8 - 271.81) \approx 0$$

(check)

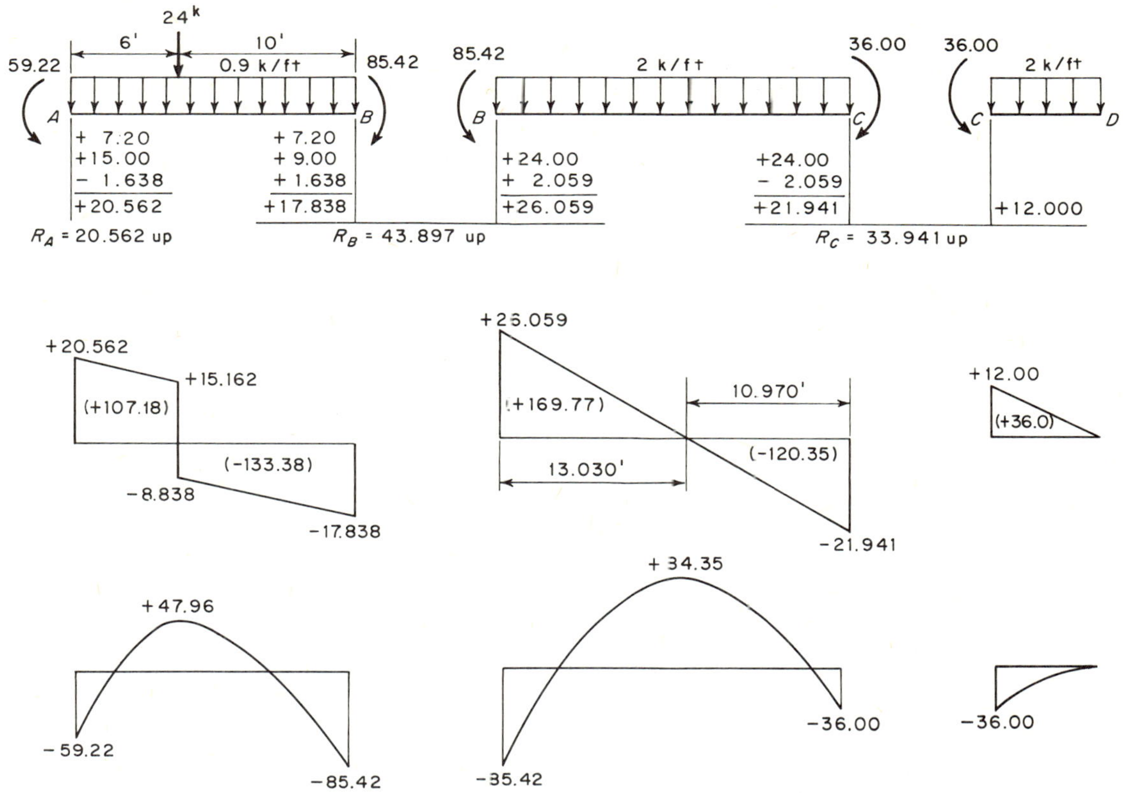

(a) Shear and moment diagrams

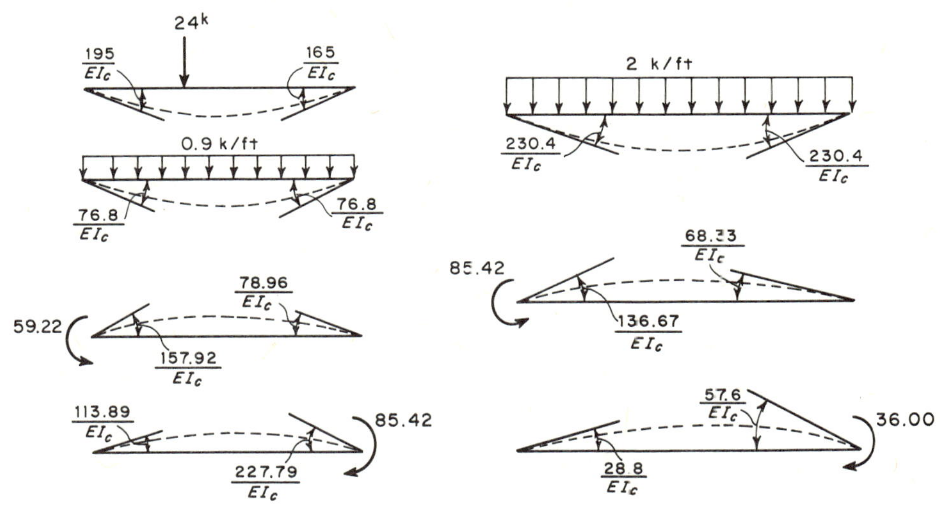

(b) Continuity checks

FIG. 6.12. Solution by the Three-Moment Equation (Example 6.3).

2) θ_B (span AB) $= \theta_B$ (span BC),

$$\frac{1}{EI_C}(-165 - 76.8 + 78.96 + 227.79) = \frac{1}{EI_C}(+23.04 - 136.67 - 28.8)$$

$$\frac{1}{EI_C}(+64.95) \approx \frac{1}{EI_C}(+64.93) \quad \text{(check)}$$

EXERCISE 6.5

Analyze the continuous beam shown in Fig. 6.13 by using the three-moment equation. Draw the shear and moment diagrams and make the continuity checks.

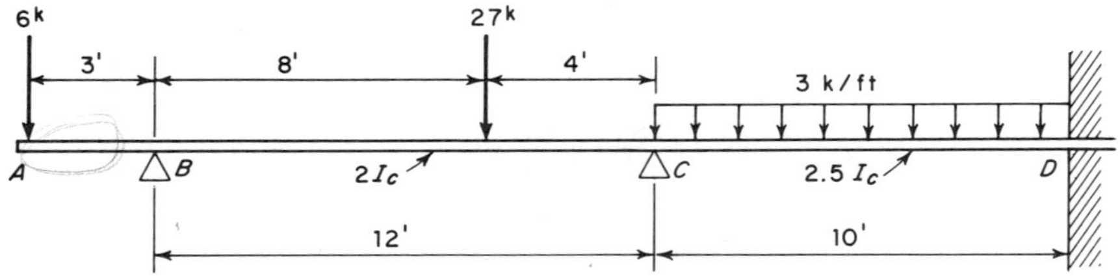

FIG. 6.13. Exercise 6.5.

6.6 Displacement Method—The Slope-Deflection Method

It has been previously stated (in Art. 6.3) that in the displacement method of analysis, the joint displacements, equal in number to the degree of freedom, are taken as the primary unknowns. The displacements of a joint in a beam may include a rotational displacement (*slope*) and a transverse displacement (*deflection*). For a typical unloaded member (or *element*) AB (Fig. 6.14a) in a beam, the four displacement quantities θ_A, Δ_A, θ_B, and Δ_B (Fig. 6.14c) define the geometry at its ends. The free-body diagram of this unloaded member is shown in Fig. 6.14d, where the four force quantities are M'_A, V'_A, M'_B, and V'_B

Had there been transverse loads applied on the typical member AB, as in Fig. 6.14b, it would have been possible to obtain the force quantities M_{0A}, R_{0A}, M_{0B}, and R_{0B} required to maintain zero slope and deflection at both ends. These fixed-condition forces are to be called *fixed-end moments and reactions* and formulas for uniform load and one concentrated load are shown in Fig. 6.14e.

By the principle of superposition, the two end moments acting on any typical member AB in a beam may be expressed as

(6.6.1a) $$M_A = M_{0A} + M'_A$$

(6.6.1b) $$M_B = M_{0B} + M'_B$$

Let ϕ_A and ϕ_B (Fig. 6.14b) be the *clockwise* end rotations measured from the

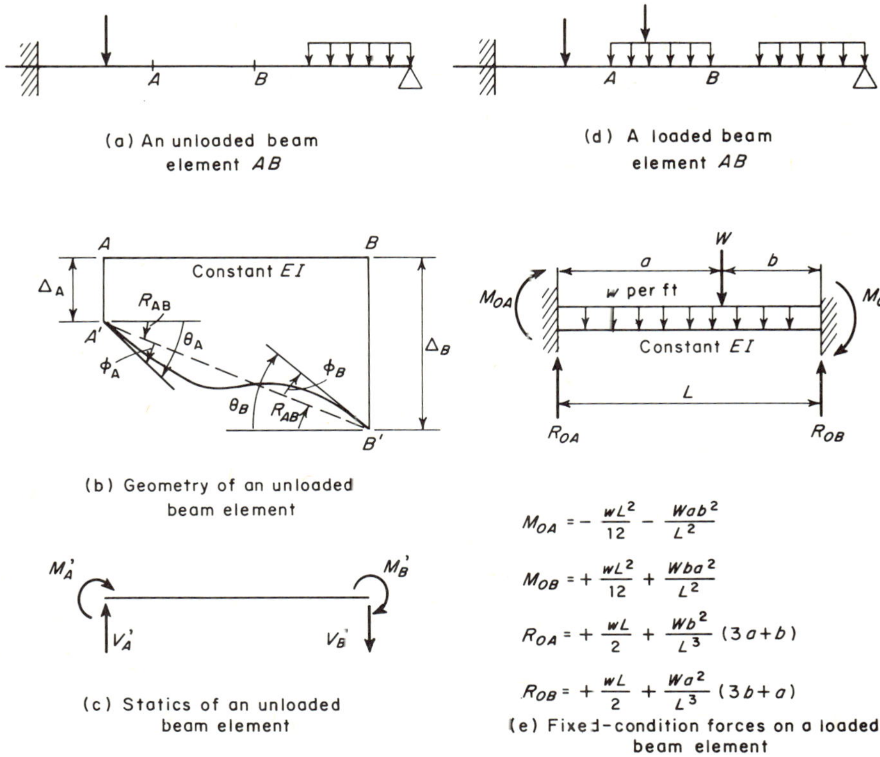

FIG. 6.14. Slope-Deflection Equations.

new member axis $A'B'$ to the tangents to the elastic curve at A and B, and let R_{AB} (Fig. 6.14b) be the *clockwise* rotation measured from the original member axis to the new member axis; then

$$R = \frac{\Delta_B - \Delta_A}{L} \tag{6.6.2}$$

and

$$\phi_A = \theta_A - R_{AB} \tag{6.6.3a}$$

$$\phi_B = \theta_B - R_{AB} \tag{6.6.3b}$$

By the conjugate-beam method,

$$\phi_A = +\frac{L}{3EI}M'_A - \frac{L}{6EI}M'_B \tag{6.6.4a}$$

$$\phi_B = -\frac{L}{6EI}M'_A + \frac{L}{3EI}M'_B \tag{6.6.4b}$$

Solving M'_A and M'_B in Eq. (6.6.4) in terms of ϕ_A and ϕ_B,

$$M'_A = \frac{4EI}{L}\phi_A + \frac{2EI}{L}\phi_B \tag{6.6.5a}$$

$$M'_B = \frac{2EI}{L}\phi_A + \frac{4EI}{L}\phi_B \tag{6.6.5b}$$

Substituting Eqs. (6.6.5) and (6.6.3) in Eq. (6.6.1),

$$M_A = M_{0A} + \frac{4EI}{L}(\theta_A - R) + \frac{2EI}{L}(\theta_B - R) \tag{6.6.6a}$$

$$M_B = M_{0B} + \frac{2EI}{L}(\theta_A - R) + \frac{4EI}{L}(\theta_B - R) \tag{6.6.6b}$$

Equations (6.6.6a and b) are called the *slope-deflection equations* because they express the clockwise end moments M_A and M_B acting on a typical member AB in a beam in terms of the applied loads on the member and the slopes and deflections at its two ends.

In the displacement method of beam analysis, the member-end moments are expressed first in terms of the unknown slopes and deflections at the joints. Then the equilibrium conditions at the joints, in one-to-one correspondence to each degree of freedom, are used to establish a system of simultaneous equations in which the joint displacements are the unknowns. Thus the degree of indeterminacy is not relevant in the slope-deflection method which can be used to analyze statically determinate as well as indeterminate beams. The degree of complexity depends on the number of joints taken by the analyst. In hand computations there is no reason to use the slope-deflection method for any statically determinate beam analysis, and there is little reason to use a degree of freedom larger than the minimum necessary in statically indeterminate beam analysis. However, in digital computation, the analyst may place joints at any locations where information on slope, deflection, shear, or moment is desired in the output. Of course, this would be at the expense of additional input data and computer time.

Example 6.4. Analyze the continuous beam of Example 6.3 (shown again in Fig. 6.15a) by the slope-deflection method.

Solution. The minimum number of joints which must be taken is three: A, B, and C, enclosing two members AB and BC. The member ends are designated as 1, 2, 3, and 4 as shown in Fig. 6.15b; the fixed-end moments are

$$M_{01} = -\frac{24(6)(10)^2}{(16)^2} - \frac{0.9(16)^2}{12} = -56.25 - 19.20 = -75.45 \text{ ft-kips}$$

$$M_{02} = +\frac{24(10)(6)^2}{(16)^2} + \frac{0.9(16)^2}{12} = +33.75 + 19.20 = +52.95 \text{ ft-kips}$$

$$M_{03} = -\frac{2(24)^2}{12} = -96.0 \text{ ft-kips}$$

$$M_{04} = +\frac{2(24)^2}{12} = +96.0 \text{ ft-kips}$$

The only freedoms possessed by the three joints are those of rotations at joints B and C; thus θ_B and θ_C are the only unknown joint displacements, and

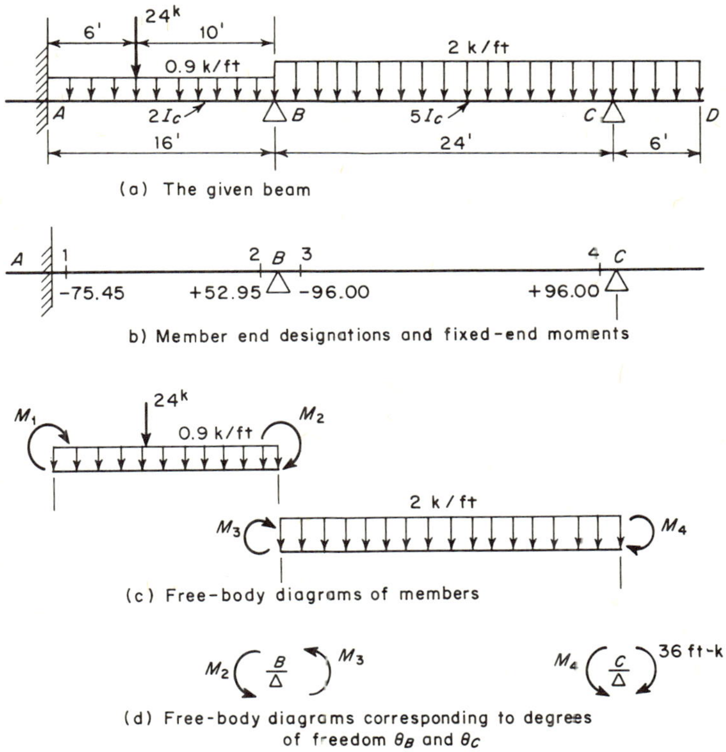

FIG. 6.15. Solution by the Slope-Deflection Method (Example 6.4).

the member axis rotations R_{AB} and R_{BC} are both equal to zero. The slope-deflection equations for the member-end moments M_1, M_2, M_3 and M_4 shown in Fig. 6.15c are

$$M_1 = -75.45 + \frac{2E(2I_c)}{16}\theta_B = -75.45 + \frac{1}{4}EI_C\theta_B$$

$$M_2 = +52.95 + \frac{4E(2I_c)}{16}\theta_B = +52.95 + \frac{1}{2}EI_C\theta_B$$

$$M_3 = -96.00 + \frac{4E(5I_c)}{24}\theta_B + \frac{2E(5I_c)}{24}\theta_C = -96.00 + \frac{5}{6}EI_c\theta_B + \frac{5}{12}EI_c\theta_C$$

$$M_4 = +96.00 + \frac{2E(5I_c)}{24}\theta_B + \frac{4E(5I_c)}{24}\theta_C = +96.00 + \frac{5}{12}EI_c\theta_B + \frac{5}{6}EI_c\theta_C$$

The equilibrium equations corresponding to the two degrees of freedom θ_B and θ_C are, from Fig. 6.15d,

$$+M_2 + M_3 = 0$$

$$+M_4 - 36.00 = 0$$

Note that since M_1, M_2, M_3, and M_4 are all unknowns, they must be considered to act clockwise on member ends and counterclockwise on the joints.

Substituting the expressions for M_1, M_2, M_3, and M_4 into the two equilibrium equations, the following two simultaneous equations are obtained:

$$\tfrac{4}{3}EI_c\theta_B + \tfrac{5}{12}EI_c\theta_C = +43.05$$

$$\tfrac{5}{12}EI_c\theta_B + \tfrac{5}{6}EI_c\theta_C = -60.00$$

Solving,

$$EI_c\theta_B = +64.93 \text{ k-ft}^2$$

$$EI_c\theta_C = -104.47 \text{ k-ft}^2$$

Substituting the known values of $EI_c\theta_B$ and $EI_c\theta_C$ into the slope-deflection equations,

$$M_1 = -75.45 + \tfrac{1}{4}(+64.93) = -59.22 \text{ ft-kips}$$

$$M_2 = +52.95 + \tfrac{1}{2}(+64.93) = +85.42 \text{ ft-kips}$$

$$M_3 = -96.00 + \tfrac{5}{6}(+64.93) + \tfrac{5}{12}(-104.47) = -85.42 \text{ ft-kips}$$

$$M_4 = +96.00 + \tfrac{5}{12}(+64.93) + \tfrac{5}{6}(-104.47) = +36.00 \text{ ft-kips}$$

The shear and moment diagrams for AB and BC may now be obtained (as in Example 6.3).

To insure the correctness of the solution, the obvious checks are for statisfaction of the equilibrium conditions equal in number to the degree of freedom. In this example they are $M_2 + M_3 = 0$ and $M_4 - 36.00 = 0$. In addition, compatibility checks equal in number to the degree of indeterminacy should be made. These are identical to the two continuity checks made in Example 6.3. One may note, however, that the compatibility checks in Example 6.3 are dependent on the simple-beam moment diagrams which were used in establishing the three-moment equations, but not in the slope-deflection equations. Thus, the compatibility checks made after usage of the slope-deflection solution would be truly independent.

Example 6.5. Using the answers to Examples 6.3 or 6.4, compute the slope and deflection at the free end of the beam shown in Fig. 6.16a. Considering the entire beam to have four joints at *A-B-C-D* and three members *AB-BC-CD*, write out the two slope-deflection equations for M_5 and M_6 in terms of θ_C-θ_D-Δ_D, and confirm that these two equations are numerically satisfied by the values of M_5, M_6, θ_C, θ_D, and Δ_D already obtained.

Solution. The value of θ_C depends only on the flexural rigidity EI of members AB and BC and has been determined in Examples 6.3 or 6.4 to be -104.47 k-ft²/(EI_c). The slope and deflection at point D, however, depend not only upon θ_C but also upon the flexural rigidity of overhang CD, whose moment of inertia is now assumed to be $3I_c$, as shown in Fig. 6.16a. By the moment-area method,

$$\theta_D = \theta_C + \left[\frac{M}{EI} \text{ area on } CD\right] = \frac{1}{EI_c}\left[-104.47 + \frac{1}{3}\left(\frac{36}{3}\right)(6)\right]$$

$$= -\frac{80.47}{EI_c} \text{ k-ft}^2$$

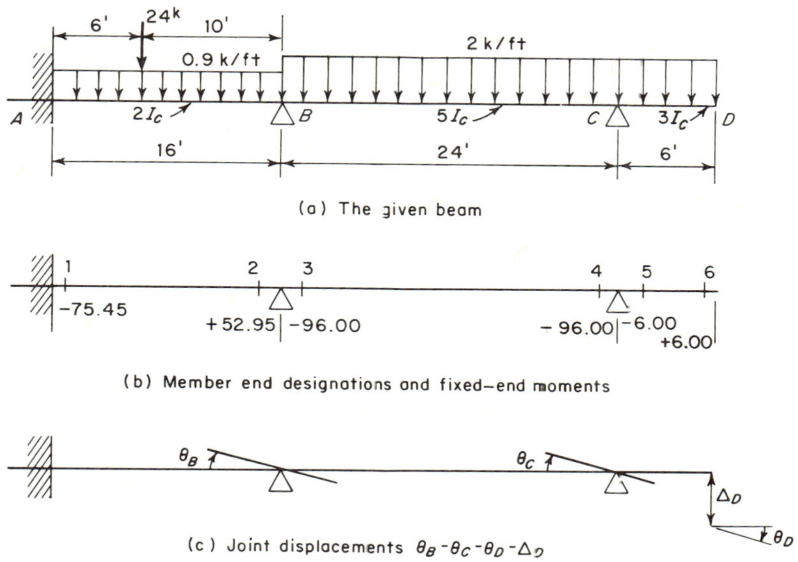

FIG. 6.16. Slope-Deflection Equations (Example 6.5).

$$\Delta_D = 6\theta_C + \left[\text{Moment of } \frac{M}{EI} \text{ area on } CD \text{ about } D\right]$$

$$= \frac{1}{EI_c}\left[6(-104.47) + \frac{1}{3}\left(\frac{36}{3}\right)(6)(4.5)\right] = -\frac{518.8}{EI_c} \text{ k-ft}^3$$

$$R_{CD} = +\frac{\Delta_D}{6} = -\frac{86.47}{EI_c} \text{ k-ft}^2$$

Note that θ_C, θ_D, and R_{CD} are positive if clockwise, and Δ_D is positive if downward.

The slope-deflection equations for end moments M_5 and M_6 acting clockwise on member CD are

$$M_5 = M_{05} + \frac{4EI}{L}(\theta_C - R_{CD}) + \frac{2EI}{L}(\theta_D - R_{CD})$$

$$= -\frac{2(6)^2}{12} + \frac{4(3)}{6}(-104.47 + 86.47) + \frac{2(3)}{6}(-80.47 + 86.47)$$

$$= -6 - 36 + 6 = -36.0 \text{ ft-kips} \quad \text{(check)}$$

$$M_6 = M_{06} + \frac{2EI}{L}(\theta_C - R_{CD}) + \frac{4EI}{L}(\theta_D - R_{CD})$$

$$= +\frac{2(6)^2}{12} + \frac{2(3)}{6}(-104.47 + 86.47) + \frac{4(3)}{6}(-80.47 + 86.47)$$

$$= +6 - 18 + 12 = 0 \quad \text{(check)}$$

EXERCISE 6.6

Analyze the continuous beam of Exercise 6.5 (Fig. 6.13) by the slope-deflection method using three joints and two members.

EXERCISE 6.7

Using the answers obtained from Exercise 6.5, compute the slope and deflection at the free end of the overhang of the continuous beam. Write out the two slope-deflection equations for the end moments acting clockwise on the overhang and confirm that these two equations are numerically satisfied. Assume that the flexural rigidity of the overhang is $1.5EI_c$.

6.7 The Statics Matrix [A] and the Deformation Matrix [B]

The continuous beam shown in Fig. 6.17a may be considered to have four joints A-B-C-D and three members AB-BC-CD. In matrix formulation it is

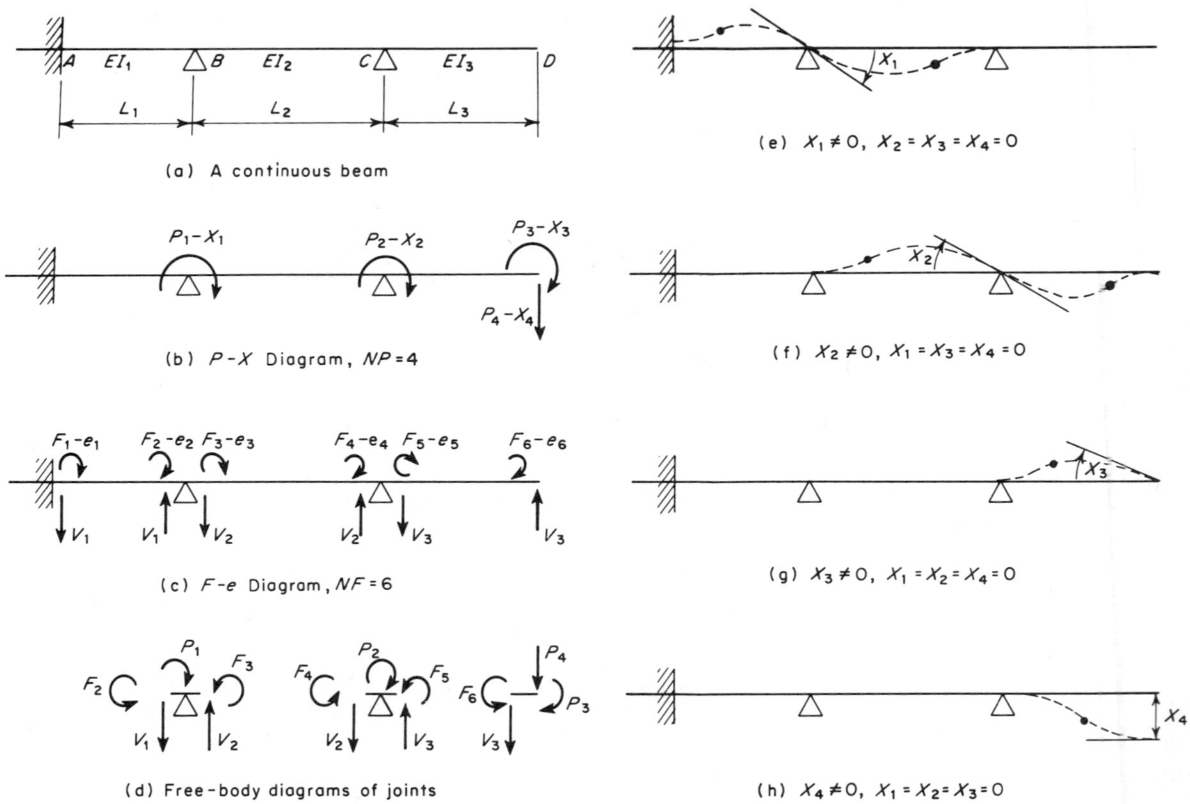

FIG. 6.17. The [A] and [B] Matrices.

convenient to label the four unknown joint displacements as X_1-X_2-X_3-X_4, instead of θ_B-θ_C-θ_D-Δ_D. Generally, the loads are applied transversely on the members between their ends. For the time being, however, assume that the loads are applied at the joints only as P_1-P_2-P_3-P_4, in one-to-one correspondence to X_1-X_2-X_3-X_4, as shown in Fig. 6.17b. The transfer of the transverse loads on the members into loads on the joints will be described later.

There is close analogy between a continuous beam and a truss, especially when the loads are applied at the joints. As in a pair of collinear pulling forces acting on the ends of a truss member, there should be a moment and a shear force acting at each end of a flexural element. The two end moments are independent, but the shear force is equal to the sum of the moments divided by the member length. Thus, the number of independent internal forces (a moment is considered to be a generalized force) in a continuous beam is twice the number of members. The end moments and shear forces acting on the member ends are shown in their positive directions in Fig. 6.17c. Once F_i and F_j are designated as the end moments, then e_i and e_j, in one-to-one correspondence, become the clockwise rotations measured from the member axis (*whether displaced or not*) to the tangents to the elastic curve at the ends.

The statics matrix $[A]$ expresses the joint forces $\{P\}$ in terms of the member end moments $\{F\}$. For instance, by observing the free-body diagrams of the joints in Fig. 6.17d,

$$P_1 = F_2 + F_3$$
$$P_2 = F_4 + F_5$$
$$P_3 = F_6$$
$$P_4 = -V_3 = -\frac{F_5 + F_6}{L_3}$$

The statics matrix $[A]$ is, therefore,

$[A]_{NP \times NF} =$

P \ F	1	2	3	4	5	6
1		+1.	+1.			
2				+1.	+1.	
3						+1.
4					$-\frac{1}{L_3}$	$-\frac{1}{L_3}$

Note that the "rectangularity" of the statics matrix, which is $(NF - NP)$, is the degree of indeterminacy of the structure.

The deformation matrix $[B]$ expresses the member end rotations $\{e\}$ in terms of the joint displacements $\{X\}$. For instance, by observing the joint displacement diagrams in Fig. 6.17e, f, g, and h,

$$e_1 = 0$$
$$e_2 = +X_1$$
$$e_3 = +X_1$$
$$e_4 = +X_2$$
$$e_5 = +X_2 - \frac{X_4}{L_3}$$
$$e_6 = +X_3 - \frac{X_4}{L_3}$$

The deformation matrix $[B]$ is, therefore,

$[B]_{NF \times NP} =$

e \ X	1	2	3	4
1				
2	+1.			
3	+1.			
4		+1.		
5		+1.		$-\frac{1}{L_3}$
6			+1.	$-\frac{1}{L_3}$

Note that the relationship $[B] = [A^T]$ always holds true. A proof by the Principle of Virtual Work may be effected for the continuous beam employing the same format as that used for the truss in Art. 4.5.

6.8 The Member Stiffness Matrix $[S]$

In the previous treatment of the slope-deflection method, the member end moments acting on an unloaded beam element have been named M'_A and M'_B; and the member end rotations, ϕ_A and ϕ_B. However, in the matrix displacement method the same member end moments are called F_i and F_j; and the same member end rotations, e_i and e_j. This apparent discrepancy arises because of the desire to unify the notations used in the matrix displacement method whether the structure is a truss, a continuous beam, or one of any other number of complex types. It suffices to declare at this time the equivalence of M'_A-M'_B to F_i-F_j, and of ϕ_A-ϕ_B to e_i-e_j. Having made this declaration, Eq. (6.6.4ab) may be expressed in matrix notation as

$$\begin{Bmatrix} e_i \\ e_j \end{Bmatrix} = [D] \begin{Bmatrix} F_i \\ F_j \end{Bmatrix}$$

in which

(6.8.1) $[D] =$

e \ F	F_i	F_j
e_i	$+\dfrac{L}{3EI}$	$-\dfrac{L}{6EI}$
e_j	$-\dfrac{L}{6EI}$	$+\dfrac{L}{3EI}$

Likewise, Eq. (6.6.5ab) may be expressed in matrix notation as

$$\begin{Bmatrix} F_i \\ F_j \end{Bmatrix} = [S] \begin{Bmatrix} e_i \\ e_j \end{Bmatrix}$$

in which

(6.8.2) $[S] =$

F \ e	e_i	e_j
F_i	$+\dfrac{4EI}{L}$	$+\dfrac{2EI}{L}$
F_j	$+\dfrac{2EI}{L}$	$+\dfrac{4EI}{L}$

The matrix $[D]$ of Eq. (6.8.1) is the member flexibility matrix of an unloaded beam element, and the matrix $[S]$ of Eq. (6.8.2) is the member stiffness matrix of an unloaded beam element.

6.9 The Joint-Force Matrix $\{P\}$

Very rarely are the loads on a continuous beam applied directly along the chosen degrees of freedom; in such cases the elements in the $\{P\}$ matrix may be easily assembled, as in truss analysis. More often the transverse loads are applied on the member lengths and these loads must be transmitted to the joints. This transmission can be effected through the fixed condition, in which all joint displacements along the degrees of freedom are zero. For instance, the given

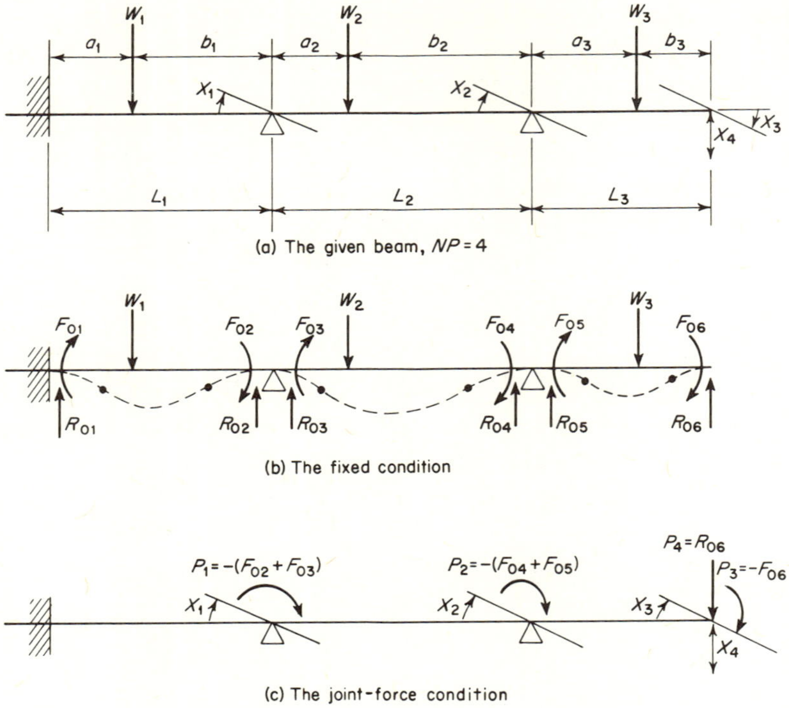

FIG. 6.18. The Joint-Force Matrix $\{P\}$.

continuous beam of Fig. 6.18a may be considered to be the sum of the fixed-conditioned beam of Fig. 6.18b and the joint-force conditioned beam of Fig. 6.18c.

When the clockwise moments $(F_{02} + F_{03})$, $(F_{04} + F_{05})$, and (F_{06}), as well as the upward force R_{06}, are to be applied to the beam of Fig. 6.18b, all joint displacements X_1 to X_4 will be zero. Then, when the reversals of these four restraining forces (a moment is considered to a generalized force) are to be applied to the joints, they will cause X_1 to X_4 in Fig. 6.18c equal in value to X_1 to X_4 in Fig. 6.18a. The total member end moments F_1^* to F_6^* in Fig. 6.18a will then be the sum of F_{01} to F_{06} in Fig. 6.18b and F_1 to F_6 in Fig. 6.18c.

The equivalent joint forces in Fig. 6.18c are

$$P_1 = -(F_{02} + F_{03}) = -\left[+\frac{W_1 b_1 a_1^2}{L_1^2} - \frac{W_2 a_2 b_2^2}{L_2^2}\right]$$

$$P_2 = -(F_{04} + F_{05}) = -\left[+\frac{W_2 b_2 a_2^2}{L_2^2} - \frac{W_3 a_3 b_3^2}{L_3^2}\right]$$

$$P_3 = -F_{06} = -\left[+\frac{W_3 b_3 a_3^2}{L_3^2}\right]$$

$$P_4 = +R_{06} = +\frac{Wa_3^2}{L_3^3}(3b_3 + a_3)$$

6.10 The Matrix-Displacement Method

The *matrix-displacement method* of continuous beam analysis involves first the establishment of the degrees of freedom NP which may be beyond the minimum number absolutely required at points of discontinuity. The number NF of independent internal forces, which are the member end moments, is equal to twice the number of members. The degree of indeterminacy is $(NF - NP)$.

Next, the method requires establishment of the statics matrix $[A]$ and the deformation matrix $[B]$, each by its own definition, with a visual check of the relationship $[B] = [A^T]$. The fixed-end moments $\{F_0\}$ are then computed and the joint-force matrix $\{P\}$ established; the elements in the $\{P\}$ matrix are the reversals of those acting on the member ends in the fixed condition.

Combining the equilibrium conditions,

$$\{P\}_{NP\times 1} = [A]_{NP\times NF}\{F\}_{NF\times 1}$$

the force-deformation relationships,

$$\{F\}_{NF\times 1} = [S]_{NF\times NF}\{e\}_{NF\times 1}$$

and the compatibility conditions,

$$\{e\}_{NF\times 1} = [B]_{NF\times NP}\{X\}_{NP\times 1}$$

the following two equations are obtained:

$$\{X\}_{NP\times 1} = [ASA^T]^{-1}_{NP\times NP}\{P\}_{NP\times 1} \tag{6.10.1}$$

and,

$$\{F\}_{NF\times 1} = [SA^T]_{NF\times NP}\{X\}_{NP\times 1} \tag{6.10.2}$$

Then, noting that the final end moments $\{F^*\}$ should be the sum of those in the fixed and joint-force conditions,

$$\{F^*\}_{NF\times 1} = \{F_0\}_{NF\times 1} + \{F\}_{NF\times 1} \tag{6.10.3}$$

The application of Eqs. (6.10.1) to (6.10.3) is illustrated in the following two examples.

Example 6.6. Using *A-B-C* as joints, analyze the continuous beam of Examples 6.3 (shown again in Fig. 5.19a) by the matrix-displacement method.

Solution. The *P–X* and *F–e* diagrams are shown in Fig. 6.19 b and c. The F_0 values shown in Fig. 6.19d are taken from Example 6.4 or Fig. 6.15b.

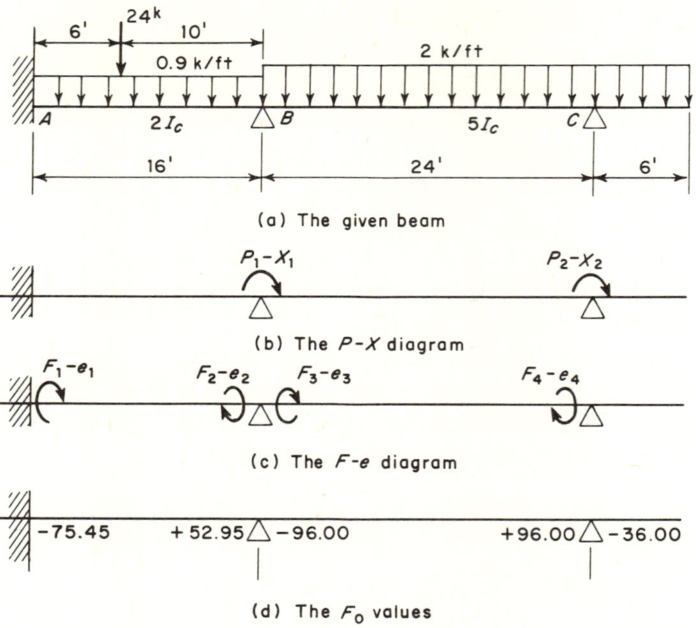

FIG. 6.19. Solution by the Matrix-Displacement Method (Example 6.6).

The $[A]$, $[B]$, $[S]$, and $\{P\}$ matrices are

$[A]_{2\times 4} =$

P \ F	1	2	3	4
1		+1.0	+1.0	
2				+1.0

$\{P\}_{2\times 1} =$

P \ LC	1
1	+43.05
2	−60.00

LC = Loading Condition

$[B]_{4\times 2} =$

e \ X	1	2
1		
2	+1.0	
3	+1.0	
4		+1.0

$[S]_{4\times 4} =$

F \ e	1	2	3	4
1	$\frac{EI_c}{2}$	$\frac{EI_c}{4}$		
2	$\frac{EI_c}{4}$	$\frac{EI_c}{2}$		
3			$\frac{5EI_c}{6}$	$\frac{5EI_c}{12}$
4			$\frac{5EI_c}{12}$	$\frac{5EI_c}{6}$

By matrix multiplication,

$$[SA^T]_{4\times 2} = \begin{array}{c|c|c|} F \backslash X & 1 & 2 \\ \hline 1 & \frac{EI_c}{4} & \\ \hline 2 & \frac{EI_c}{2} & \\ \hline 3 & \frac{5EI_c}{6} & \frac{5EI_c}{12} \\ \hline 4 & \frac{5EI_c}{12} & \frac{5EI_c}{6} \end{array} \qquad [ASA^T]_{2\times 2} = \begin{array}{c|c|c|} P \backslash X & 1 & 2 \\ \hline 1 & \frac{4EI_c}{3} & \frac{5EI_c}{12} \\ \hline 2 & \frac{5EI_c}{12} & \frac{5EI_c}{6} \end{array}$$

Rather than solving for $\{X\}$ from $\{X\} = [ASA^T]^{-1}\{P\}$, the values of X_1 and X_2 are taken from those of θ_B and θ_C in the slope-deflection solution of Example 6.4. A check is made to see that the equation $\{P\} = [ASA^T]\{X\}$ is satisfied. Thus,

$$\begin{bmatrix} +43.05 \\ -60.00 \end{bmatrix} = \begin{bmatrix} \frac{4EI_c}{3} & \frac{5EI_c}{12} \\ \frac{5EI_c}{12} & \frac{5EI_c}{6} \end{bmatrix} * \begin{bmatrix} +\frac{64.93}{EI_c} \\ -\frac{104.47}{EI_c} \end{bmatrix}$$

$$\{F^*\} = \begin{bmatrix} F_1^* \\ F_2^* \\ F_3^* \\ F_4^* \end{bmatrix} = \begin{bmatrix} F_{01} \\ F_{02} \\ F_{03} \\ F_{04} \end{bmatrix} + [SA^T] \begin{bmatrix} X_1 \\ X_2 \end{bmatrix}$$

$$= \begin{bmatrix} -75.45 \\ +52.95 \\ -96.00 \\ +96.00 \end{bmatrix} + \begin{bmatrix} \frac{EI_c}{4} & \\ \frac{EI_c}{2} & \\ \frac{5EI_c}{6} & \frac{5EI_c}{12} \\ \frac{5EI_c}{12} & \frac{5EI_c}{6} \end{bmatrix} * \begin{bmatrix} +\frac{64.93}{EI_c} \\ -\frac{104.47}{EI_c} \end{bmatrix}$$

$$
= \begin{bmatrix} -75.45 \\ +52.95 \\ -96.00 \\ +96.00 \end{bmatrix} + \begin{bmatrix} +16.23 \\ +32.47 \\ +10.58 \\ -60.00 \end{bmatrix} = \begin{bmatrix} -59.22 \\ +85.42 \\ -85.42 \\ +36.00 \end{bmatrix}
$$

In comparing the above solution with that of the slope-deflection method in Example 6.4, one will find that the slope-deflection equations, when cast in matrix form, are in fact $\{F\}^* = \{F_0\} + [SA^T]\{X\}$, and the simultaneous equations in the slope-deflection method are $\{P\} = [ASA^T]\{X\}$.

The matrix-displacement method, when displayed and carried out in longhand, may actually take more time than the conventional slope-deflection method; however, the former method can be conveniently programmed on the computer by taking $[A]$, $[S]$, and $\{P\}$ as input, and $\{X\}$ and $\{F\}$ as output. When there are several loading conditions, the P-matrix will simply become one with several columns.

Example 6.7. Using A-B-C-D as joints, analyze the continuous beam of Example 6.5, shown again in Fig. 6.20a, by the matrix displacement method. Rather than solving the equation $\{P\} = [ASA^T]\{X\}$, note that this equation is satisfied by the values of θ_B-θ_C-θ_D-Δ_D already obtained in Example 6.5

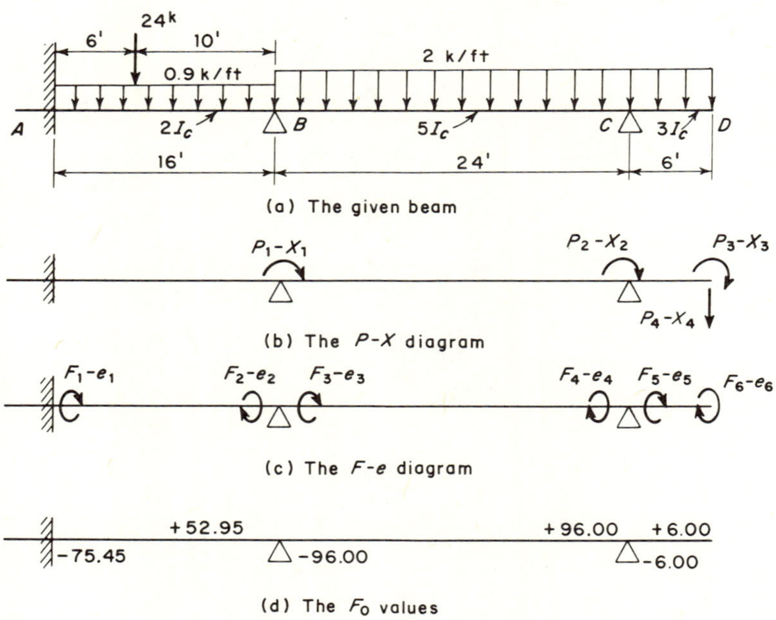

FIG. 6.20. Solution by the Matrix-Displacement Method (Example 6.7).

Solution. The P–X and F–e diagrams are shown in Fig. 6.20b and c and the F_0 values shown in Fig. 6.20d are taken from Example 6.5 or Fig. 6.16b. The $[A]$, $[B]$, $[S]$, and $\{P\}$ matrices are

$[A]_{4\times 6} =$

P \ F	1	2	3	4	5	6
1		+1.0	+1.0			
2				+1.0	+1.0	
3						+1.0
4					$-\frac{1}{6}$	$-\frac{1}{6}$

$\{P\} =$

P \ LC	1
1	+43.05
2	−90.00
3	− 6.00
4	+ 6.00

$[B]_{6\times 4} =$

e \ X	1	2	3	4
1				
2	+1.0			
3	+1.0			
4		+1.0		
5		+1.0		$-\frac{1}{6}$
6			+1.0	$-\frac{1}{6}$

$[S]_{6\times 6} = EI_c *$

F \ e	1	2	3	4	5	6
1	$\frac{1}{2}$	$\frac{1}{4}$				
2	$\frac{1}{4}$	$\frac{1}{2}$				
3			$\frac{5}{6}$	$\frac{5}{12}$		
4			$\frac{5}{12}$	$\frac{5}{6}$		
5					2	1
6					1	2

By matrix multiplication,

$[SA^T]_{6\times 4} = EI_c *$

F \ X	1	2	3	4
1	$\frac{1}{4}$			
2	$\frac{1}{2}$			
3	$\frac{5}{6}$	$\frac{5}{12}$		
4	$\frac{5}{12}$	$\frac{5}{6}$		
5		2	1	$-\frac{1}{2}$
6		1	2	$-\frac{1}{2}$

$[ASA^T]_{4\times 4} = EI_c *$

P \ X	1	2	3	4
1	$\frac{4}{3}$	$\frac{5}{12}$		
2	$\frac{5}{12}$	$\frac{17}{6}$	1	$-\frac{1}{2}$
3		1	2	$-\frac{1}{2}$
4		$-\frac{1}{2}$	$-\frac{1}{2}$	$+\frac{1}{6}$

Note that the $[ASA^T]$ matrix is always symmetric, as it has been proved previously in Art. 4.5. From Example 6.5,

$\{X\}_{4\times 1} =$

X \ LC	1
1	$+\frac{64.93}{EI_c}$
2	$-\frac{104.47}{EI_c}$
3	$-\frac{80.47}{EI_c}$
4	$-\frac{518.8}{EI_c}$

The equation $\{P\} = [ASA^T]\{X\}$ is numerically checked out as shown below:

$$\begin{Bmatrix} +43.05 \\ -90.00 \\ -\ 6.00 \\ +\ 6.00 \end{Bmatrix} = \begin{bmatrix} \frac{4}{3} & \frac{5}{12} & & \\ \frac{5}{12} & \frac{17}{6} & 1 & -\frac{1}{2} \\ & 1 & 2 & -\frac{1}{2} \\ & -\frac{1}{2} & -\frac{1}{2} & +\frac{1}{6} \end{bmatrix} * \begin{Bmatrix} +\ 64.93 \\ -104.47 \\ -\ 80.47 \\ -518.8 \end{Bmatrix}$$

$$
\{F^*\} = \begin{Bmatrix} -75.45 \\ +52.95 \\ -96.00 \\ +96.00 \\ -\ 6.00 \\ +\ 6.00 \end{Bmatrix} + \begin{bmatrix} \frac{1}{4} & & & \\ \frac{1}{2} & & & \\ \frac{5}{6} & \frac{5}{12} & & \\ \frac{5}{12} & \frac{5}{6} & & \\ & 2 & 1 & -\frac{1}{2} \\ & 1 & 2 & -\frac{1}{2} \end{bmatrix} * \begin{Bmatrix} +\ 64.93 \\ -104.47 \\ -\ 80.47 \\ -518.80 \end{Bmatrix}
$$

$$
= \begin{Bmatrix} -75.45 \\ +52.95 \\ -96.00 \\ +96.00 \\ -\ 6.00 \\ +\ 6.00 \end{Bmatrix} + \begin{Bmatrix} +16.23 \\ +32.47 \\ +10.58 \\ -60.00 \\ -30.00 \\ -\ 6.00 \end{Bmatrix} = \begin{Bmatrix} -59.22 \\ +85.42 \\ -85.42 \\ +36.00 \\ -36.00 \\ 0.00 \end{Bmatrix}
$$

EXERCISE 6.8

Analyze the continuous beam of Exercise 6.5 (Fig. 6.13) by the matrix-displacement method using three joints and two members.

EXERCISE 6.9

Using the free end of the overhang as one of the joints, analyze the continuous beam of Exercise 6.5 (Fig. 6.13) by the matrix-displacement method. As in Exercise 6.7, assume that the flexural rigidity of the overhang is $1.5EI_c$.

6.11 The Moment-Distribution Method

The moment-distribution method was originated by Professor Hardy Cross in the 1930's and has been the most widely used method in continuous beam and rigid frame analysis. Its chief advantage lies in its enabling the analyst to obtain a quick and efficient hand solution as well as a thorough understanding of the physical meaning in the computation processes. Essentially, this method involves the solution of the simultaneous equations generated in either the slope-deflection method or the matrix-displacement method by a relaxation procedure.

In its elementary version the moment-distribution method applied only to cases in which all degrees of freedom are in rotation. It begins with the fixed condition and allows the joint rotations to occur in graduated steps until moment equilibrium at each joint is satisfied to a prescribed tolerance. Consider, for instance, the effect of a joint rotation θ_A at a typical interior support of a continuous beam as shown in Fig. 6.21. The near-end moments F_2 and F_3 are

$$F_2 = \frac{4EI_{1\text{-}2}}{L_{1\text{-}2}}\theta_A \qquad F_3 = \frac{4EI_{3\text{-}4}}{L_{3\text{-}4}}\theta_A$$

in which the quantity $4EI/L$ is known as the *stiffness factor*. Thus the stiffness factor is the ratio of the moment at the near end to the rotation at that end when the far end is fixed. The far-end moments F_1 and F_4 are

$$F_1 = \tfrac{1}{2}F_2 \qquad F_4 = \tfrac{1}{2}F_3$$

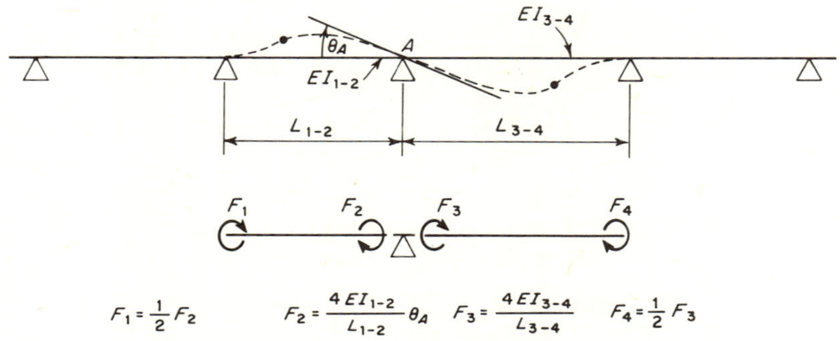

FIG. 6.21. Stiffness and Carry-Over Factors.

in which the quantity $\frac{1}{2}$ is known as the *carry-over factor*. Thus the carry-over factor is the ratio of the far-end moment to the near-end moment. The *distribution factor* at each of the two near ends meeting at a common support is defined as the ratio of each stiffness factor to the sum of the two stiffness factors.

In the moment distribution process the fixed end moments at all member ends are computed first. Then a joint rotation is allowed at each degree of freedom, of such a magnitude as to make the sum of the member end moments at that joint equal to zero. In the meantime, two carry-over moments, each equal to one-half of the balancing moments added at the near end, are placed at the far ends and treated as new fixed end moments. Note that the distribution factor at an exterior simple support (with or without an overhang) is one, and the name distribution factor at a fixed support has no meaning because there can be no rotation there.

The use of the moment-distribution method in continuous beam analysis will be illustrated by the following example.

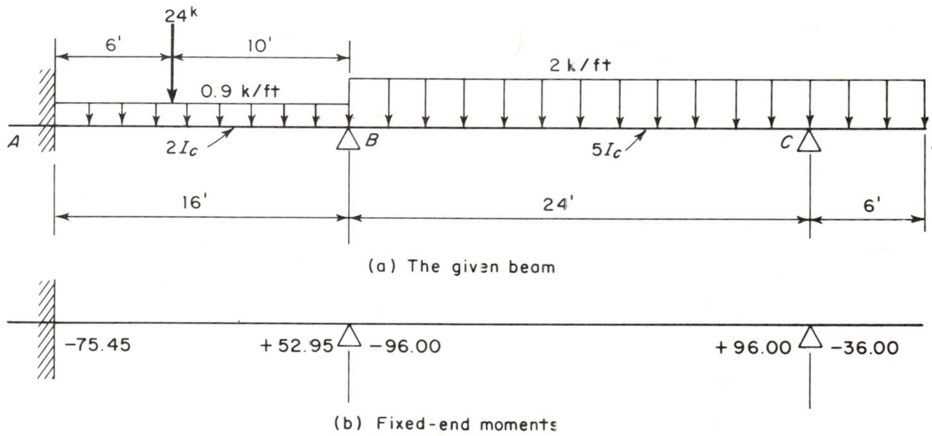

FIG. 6.22. Solution by the Moment-Distribution Method (Example 6.8).

Example 6.8. Analyze the continuous beam of Example 6.6, shown again in Fig. 6.22a, by the moment-distribution method.

Solution. The required moment distribution is set up in a convenient tabular form as shown in Table 6.1. The overhang is only a means of applying a joint moment at C and as such is not a member in this two-degree-of-freedom problem. The distribution factors are proportional to the stiffness factors $4EI/L$, and also to EI/L. Since joint A is fixed and needs no balancing, the name distribution factor does not apply and a dash is placed under AB in Table 6.1. The distribution factors at joint B are $0.1250/(0.1250 + 0.2083) = 0.3750$ to BA and $0.2083/(0.1250 + 0.2083) = 0.6250$ to BC. Because only one member enters joint C, the distribution factor is 1.0000 to CB.

The fixed-end moments have been computed previously in Examples 6.4 and 6.6; they are shown again in Fig. 6.22b and entered in Table 6.1. Note that it is decided at the outset that this moment distribution is to be carried to two decimal places.

In the first-cycle balance, a dash is placed under AB because end A of member AB can have any moment assigned to it. The unbalance at joint B is $(+52.95 - 96.00) = -43.05$; so the balancing moments are both positive. These are the products of the distribution factors and 43.05. The unbalance at joint C is $+60.00$; so the balancing moment at end C of member CB is -60.00.

The carry-overs at the beginning of the second cycle are one-half of the balancing moments at the far ends in the first cycle; and are of the same sign. In splitting an odd number, the usual practice of making the last digit even is adopted. Note that the moment distribution comes to a stop at the end of the tenth cycle where the balancing moments become numerically equal to 0.01. Then the total moments are the sums of the moments in the respective columns. These values check well with the results previously obtained by the slope-

Table 6.1. Moment Distribution (Example 6.8)

Joint		A	B		C	
Member		AB	BA	BC	CB	Overhang
EI/L		$0.1250EI_c$	$0.1250EI_c$	$0.2083EI_c$	$0.2083EI_c$	
D.F.		——	0.3750	0.6250	1.0000	
1	FEM	−75.45	+52.95	−96.00	+96.00	−36.00
	BAL	——	+16.14	+26.91	−60.00	
2	CO	+ 8.07	——	−30.00	+13.46	
	BAL	——	+11.25	+18.75	−13.46	
3	CO	+ 5.62	——	− 6.73	+ 9.38	
	BAL	——	+ 2.52	+ 4.21	− 9.38	
4	CO	+ 1.26	——	− 4.69	+ 2.10	
	BAL	——	+ 1.76	+ 2.93	− 2.10	
5	CO	+ 0.88	——	− 1.05	+ 1.46	
	BAL	——	+ 0.39	+ 0.66	− 1.46	
6	CO	+ 0.20	——	− 0.73	+ 0.33	
	BAL	——	+ 0.27	+ 0.46	− 0.33	
7	CO	+ 0.14	——	− 0.16	+ 0.23	
	BAL	——	+ 0.06	+ 0.10	− 0.23	
8	CO	+ 0.03	——	− 0.12	+ 0.05	
	BAL	——	+ 0.04	+ 0.08	− 0.05	
9	CO	+ 0.02	——	− 0.02	+ 0.04	
	BAL	——	+ 0.01	+ 0.01	− 0.04	
10	CO	——	——	− 0.02	——	
	BAL	——	+ 0.01	+ 0.01	——	
Total		−59.23	+85.40	−85.40	+36.00	−36.00
Check	Change	+16.22	+32.45	+10.60	−60.00	
	$-\frac{1}{2}$ Change	−16.22	− 8.11	+30.00	− 5.30	
	Sum	0.00	+24.34	+40.60	−65.30	
	$\frac{3EI}{L}$	$0.375EI_c$	$0.375EI_c$	$0.625EI_c$	$0.625EI_c$	
	$\theta = \frac{\text{Sum}}{3EI/L}$	0.00	$+\frac{64.91}{EI_c}$	$+\frac{64.96}{EI_c}$	$-\frac{104.48}{EI_c}$	

deflection or matrix-displacement methods; the largest difference might be as much as ±0.02 or ±0.03, but should not be any larger.

6.12 Check on Moment Distribution

When member axis does not rotate ($R = 0$), the slope-deflection equations (6.6.6a and b) become

(6.12.1a) $$M_A = M_{0A} + \frac{4EI}{L}\theta_A + \frac{2EI}{L}\theta_B$$

(6.12.1b) $$M_B = M_{0B} + \frac{2EI}{L}\theta_A + \frac{4EI}{L}\theta_B$$

Solving the above equations for θ_A and θ_B,

(6.12.2a) $$\theta_A = \frac{(M_A - M_{0A}) - \frac{1}{2}(M_B - M_{0B})}{(3EI/L)}$$

(6.12.2b) $$\theta_B = \frac{(M_B - M_{0B}) - \frac{1}{2}(M_A - M_{0A})}{(3EI/L)}$$

Equations (6.12.2a and b) may be used to retrieve the member end rotations from the member end-moment values in the moment distribution table, by asserting that the member end rotation at the near end is equal to the sum of the change in moment at the near end and negative one-half of the change at the far end, divided by $3EI/L$. When this is done, as in the check at the bottom of Table 6.1, the value of θ_A comes out to be zero as expected, and the same value of θ_B is obtained under BA and BC. These two compatibility checks indicate that the moment distribution has been correctly performed on the basis of the fixed end moments and the distribution factors listed at the beginning of Table 6.1.

6.13 Modified Stiffness Factor and Modified Fixed-End Moment

When the exterior support of a continuous beam is a simple support, either with or without an overhang, it may be treated as such at the outset of moment distribution by modifying both the stiffness factor and the fixed-end moment at the interior end of the exterior span. When this is done, the exterior joint needs no further treatment during the entire moment distribution process.

The stiffness factor $S = 4EI/L$ has been defined in Art. 6.11 as being the

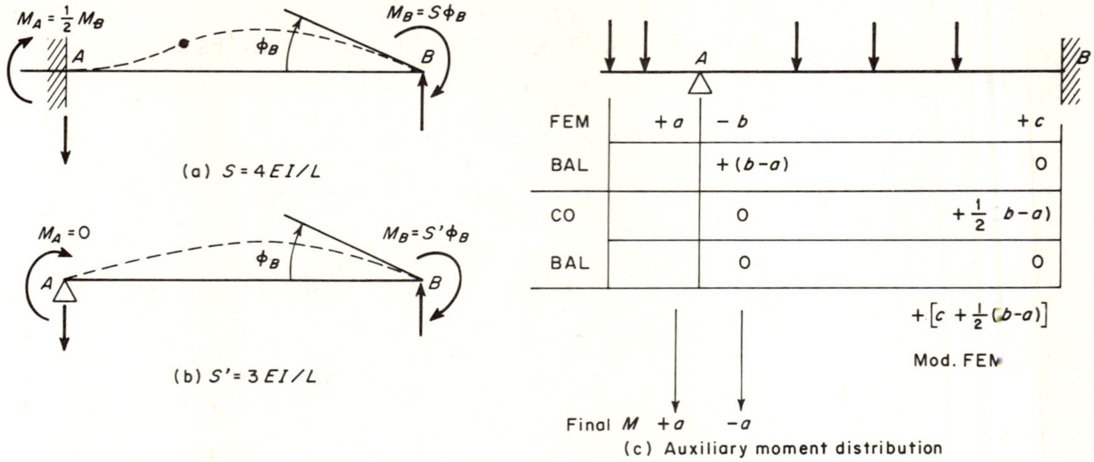

FIG. 6.23. Modified Stiffness Factor and *FEM*.

ratio of the moment at the near end (end B in Fig. 6.23a) to the rotation at that end when the far end (end A in Fig. 6.23a) is fixed. When the far end (end A in Fig. 6.23b) is a simple support, the ratio of the moment at the near end (end B in Fig. 6.23b) to the rotation at that end may be conveniently derived by the conjugate-beam method to be $S' = 3EI/L$. Therefore, the modified stiffness factor S' is $\frac{3}{4}$ of the regular stiffness factor S.

The modified fixed-end moment at end B of an overhanging beam AB may be obtained by performing the auxiliary moment distribution shown in Fig. 6.23c.

By modifying both the stiffness factor and the fixed-end moment at the adjacent interior end of an exterior simple support, the moment distribution for the entire continuous beam may be simplified somewhat because neither carry-over nor balance will ever appear further at the exterior support beyond the auxiliary moment distribution. In the following numerical example, in which only one joint remains to be balanced after the exterior support is treated as a simple support at the outset, only two cycles are needed to bring the moment distribution to an automatic close.

Example 6.9. Rearrange the moment distribution done in Example 6.8 by modifying the stiffness factor at end B of member BC.

Solution. Referring to Fig. 6.22 and Table 6.2, the modified distribution factors at joint B are computed on the basis of the regular EI/L value of member BA and the reduced $0.75EI/L$ value of member BC. Joint C is being treated as a simple support in the entire moment distribution process; thus, the name distribution factor does not apply any more and a dash is placed under CB in Table 6.2.

The auxiliary moment distribution is performed on the overhanging beam BCD which has a fixed support at B and a simple support at C. In Table 6.2

the moment distribution comes to an automatic close at the end of the second cycle. The check on moment distribution should be performed exactly as in Table 6.1.

Table 6.2. Moment Distribution (Example 6.9)

Joint	*A*	*B*		*C*	
Member	*AB*	*BA*	*BC*	*CB*	Overhang
$\frac{EI}{L}$	$0.1250EI_c$	$0.1250EI_c$	$0.2083EI_c$	$0.2083EI_c$	
$0.75\frac{EI}{L}$			$0.1562EI_c$		
Mod. D. F.	——	0.4444	0.5556	——	
FEM	−75.45	+52.95	−96.00 −30.00	+96.00 −60.00	−36.00
1 Mod. *FEM* BAL	−75.45 ——	+52.95 +32.46	−126.00 +40.59	+36.00	−36.00
2 CO BAL	+16.23 ——	—— ——	—— ——		
Total	−59.22	+85.41	−85.41	+36.00	−36.00

The check on moment distribution is the same as in Table 6.1.

EXERCISE 6.10

Analyze the continuous beam of Exercise 6.5 (Fig. 6.13) by the moment-distribution method using the regular stiffness of member *CB* at joint *C*.

EXERCISE 6.11

Analyze the continuous beam of Exercise 6.5 (Fig. 6.13) by the moment-distribution method using the modified stiffness of member *CB* at joint *C*.

6.14 Concluding Remarks

The force and displacement methods of analyzing continuous beams have been treated in parallel. Both methods deserve attention for many good reasons. In particular, the Theorem of Least Work (presented later in Chapter 9) will be shown to be the theoretical equivalent of the force method.

The matrix-displacement method, when programmed on the computer, is useful in cases where there are numerous spans in the continuous beam and numerous loading conditions are to be analyzed.

The conventional slope-deflection method furnishes the bridge toward the understanding of the matrix displacement method on one side and the moment-distribution method on the other side.

The moment-distribution method may continue to be useful in preliminary analysis and design, wherein some subassemblies of a complex continuous structure may be studied by simply carrying out two or three cycles of moment distribution.

Rigid Frames

7

7.1 Definitions and Fundamental Assumptions

A *rigid frame*, in the most general sense, is a skeletal structure wherein the constituent members meet at rigid joints, as typified by welded joints in structural steel or monolithically poured joints in reinforced concrete. Treated in this chapter, however, are only plane rigid frames which are composed of straight prismatic members and subjected to loads in the plane of the frame itself.

Very often, two rigid frames, one on each edge of a highway bridge, are designed to support the floor beams transverse to the direction of traffic. Evenly spaced rigid frames may be used to support the floor beams in a building or the purlins under a roof. Assemblies of complex spatial rigid frames may be involved in common structures like bridges and buildings or in special structures like flight vehicles and ocean vessels.

Plane rigid frames, being dealt with here, may be either statically determinate or indeterminate. Although statically determinate rigid frames, as shown in Fig. 7.1a, b and c, frequently might not be built as such in practice, their analysis is usually prerequisite in the force method of analyzing indeterminate rigid frames. In fact, the rigid frames in Fig. 7.1a, b and c are the degenerate structures of those in Fig. 7.1d, e and f, respectively.

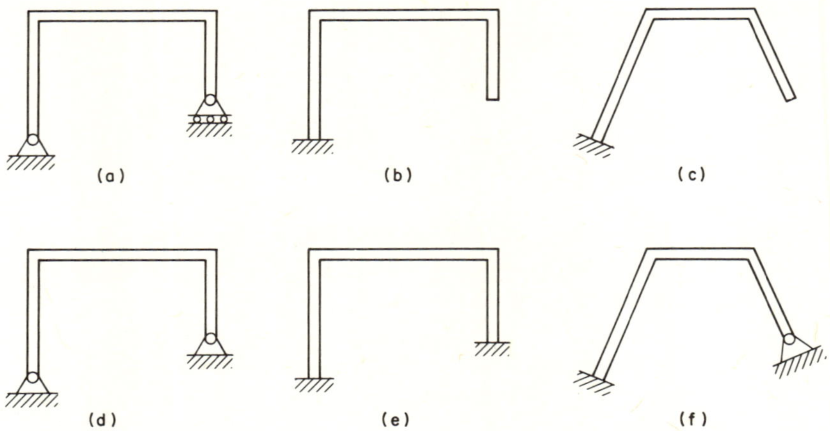

FIG. 7.1. Statically Determinate and Indeterminate Rigid Frames.

A straight member in a rigid frame, such as one of those in Fig. 7.1, should be subjected to axial force, shear, and bending moment along its length. Treatment of shear deformation is to be deferred since its effect is usually quite small in flexural members of ordinary proportions. Axial deformation is the only kind which exists in the members of a truss and has been duly considered in truss analysis. Its effect, however, in the rotations and deflections of the joints in a rigid frame is overshadowed by that of the bending moment; in fact, its

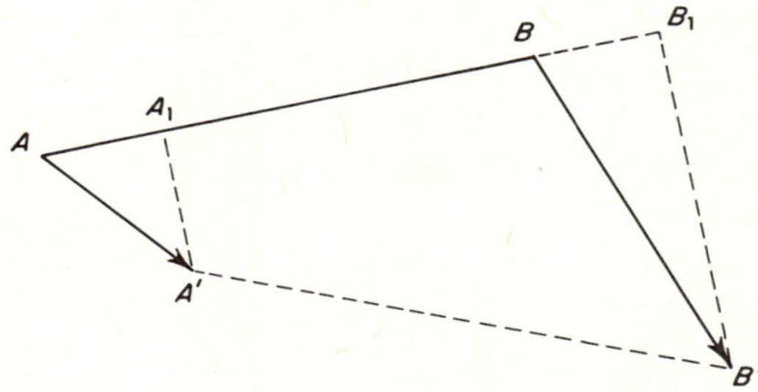

FIG. 7.2. Plane Motion of a Member in a Rigid Frame.

effect on the magnitudes of the redundant forces in an indeterminate rigid frame is usually on the order of several percent. Thus, the discussion in this chapter will be confined to the assumption that the axial forces in the members of a rigid frame do not cause changes in member lengths.

In first-order analysis, the transverse displacement of either end of a member in a structure is assumed not to change its length. When the member length does not change, the longitudinal components of the end displacements, such as AA_1 and BB_1 in Fig. 7.2, must be equal. In this way, the new member length $A'B'$ is considered to be equal to the original member length AB.

The two fundamental assumptions just discussed, that member lengths are not changed by axial forces and transverse displacements, should be *continuously* kept in mind during the study of the rest of this chapter.

7.2 Rotations and Deflections of the Joints in a Rigid Frame

The rotations and deflections of the joints in a rigid frame may be determined by either the moment-area/conjugate-beam method or the unit-load method, provided the bending moments along the lengths of all the constituent members are known, as in the case of statically determinate rigid frames.

The moment-area/conjugate-beam theorems can be applied to a straight member in a rigid frame in the same manner as to a beam. The procedure involves the following important steps:

1) Obtain the reactions and sketch the bending moment diagram for the entire statically determinate rigid frame; 2) locate the probable new positions of the joints as well as their rotations, and sketch an elastic curve consistent with the bending moment diagram; 3) compute the modified moment areas to take account of the variable flexural rigidity of each member; and 4) apply the moment-area/conjugate-beam theorems in succession between the ends of each straight member to arrive at the positions and rotations of the joints.

In the unit-load method of finding the rotations and deflections of the joints in a rigid frame, the working formulas (5.7.6) and (5.7.7)

$$P_i * Y_i = \int \frac{Mmdx}{EI}$$

$$P_i * \theta_i = \int \frac{Mmdx}{EI}$$

should include the integration along all the members in the structure. The reader only needs to review the derivation presented in Art. 5.7 and keep in mind that the Principle of Virtual Work is now applied to the entire rigid frame instead of a beam.

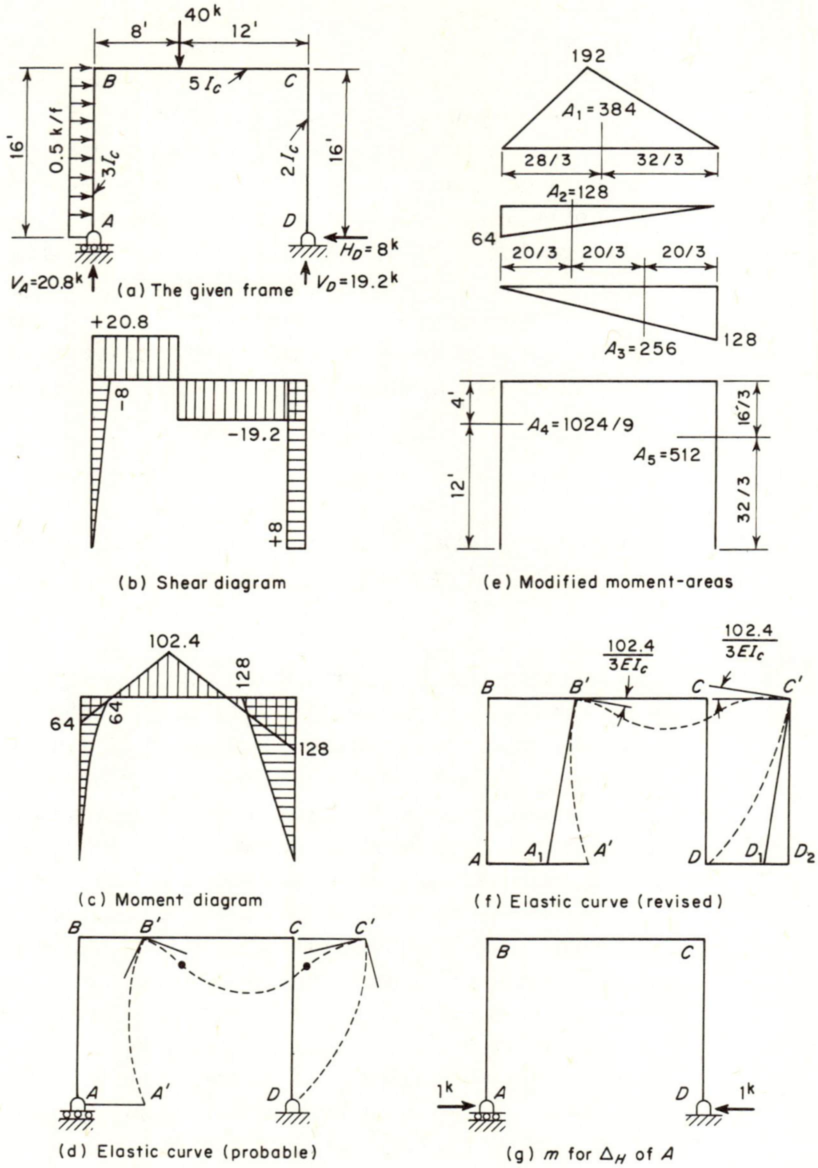

FIG. 7.3. The Rigid Frame of Example 7.1.

Example 7.1. Determine the horizontal deflection of the roller support at A of the statically determinate rigid frame $ABCD$ shown in Fig. 7.3a, first by the moment-area/conjugate-beam method; then by the unit-load method.

Solution

a) *Compute reactions and sketch the moment diagram.* Taking moments about A in Fig. 7.3a,

$$40(8) + 8(8) = 20V_D$$

$$V_D = 19.2 \text{ kips}$$

Taking moments about D in Fig. 7.3a,

$$40(12) = 8(8) + 20V_A$$
$$V_A = 20.8 \text{ kips}$$

Equating to zero the sum of horizontal forces in Fig. 7.3a,

$$H_D = 8 \text{ kips}$$

The free-body, shear, and moment diagrams of each member may be investigated and determined, one member at a time. To conserve space here, however, the composite shear and moment diagrams are shown in Fig. 7.3b and c. The usual sign convention is applied to the shear values in Fig. 7.3b by viewing the vertical member from the right-hand side; but signs are omitted before the moment values in Fig. 7.3c, with the understanding that the moment diagram is always plotted on the compression side of the member.

b) *Sketch the elastic curve.* The elastic curve of the entire rigid frame is qualitatively shown in Fig. 7.3d. The lateral load would shift the horizontal member BC to $B'C'$, with $BB' = CC'$. The heavy vertical load would probably (to be confirmed later) rotate the joint at B' in the clockwise direction and the joint at C' in the counterclockwise direction. The elastic curves $B'C'$, $A'B'$, and $C'D'$, are then sketched in, noting that the compression side should be the concave side.

c) *Compute the modified moment areas.* The moment diagram of member BC is separated into three parts: one is due to the vertical load acting on a simple beam, and the remaining two are due to each of the two end moments. These three parts are shown in Fig. 7.3e. The modified moment-areas A_1 to A_5, with their centroidal locations, shown in Fig. 7.3e, are

$$A_1 = \tfrac{1}{2}(\tfrac{192}{5})(20) = 384 \qquad A_2 = \tfrac{1}{2}(\tfrac{64}{5})(20) = 128$$
$$A_3 = \tfrac{1}{2}(\tfrac{128}{5})(20) = 256 \qquad A_4 = \tfrac{1}{3}(\tfrac{64}{3})(16) = \tfrac{1024}{9}$$
$$A_5 = \tfrac{1}{2}(\tfrac{128}{2})(16) = 512$$

d) *Apply the moment-area/conjugate-beam theorems.* Applying Conjugate-Beam Theorem I to $B'C'$

$$EI_c\theta_B = \frac{A_1(\frac{32}{3})}{20} - \tfrac{2}{3}A_2 - \tfrac{1}{3}A_3 = \tfrac{1}{3}(614.4 - 256 - 256) = \frac{102.4}{3}$$

$$EI_c\theta_C = -\frac{A_1(\frac{28}{3})}{20} + \tfrac{1}{3}A_2 + \tfrac{2}{3}A_3 = \tfrac{1}{3}(-537.6 + 128 + 512) = \frac{102.4}{3}$$

Since the above results show both θ_B and θ_C to be clockwise, a revised elastic curve is shown in Fig. 7.3f. Applying Moment-Area Theorem II to $C'D$,

$$DD_2 = DD_1 + D_1D_2 = \frac{A_5}{EI_c}\left(\frac{32}{3}\right) + 16\theta_C = \frac{1}{3EI_c}(16384 + 1638.4)$$
$$= \frac{18022.4}{3EI_c} = CC' = BB'$$

Applying Moment-Area Theorem II to $A'B'$,

$$AA' = AA_1 + A_1A' = (BB' - 16\theta_B) + A_1A'$$
$$= \frac{1}{EI_c}\left[\frac{18022.4}{3} - \frac{16(102.4)}{3} + 12A_4\right]$$
$$= \frac{1}{3}EI_c\,(18022.4 - 1638.4 + 4096) = \frac{20480}{3EI_c}\text{ k-ft}^3$$

e) *Compute Δ_H of A by the unit-load method.* The M and m values are observed from Fig. 7.3a and g and used in the working formula of the unit-load method.

$$EI_c\Delta_H \text{ of } A = \int \frac{Mmdx}{EI} = \tfrac{1}{3}\int_0^{16}\left(-\frac{0.5x^2}{2}\right)(-x)dx + \tfrac{1}{5}\int_0^{8}(-64 + 20.8x)(16)\,dx$$
$$+ \tfrac{1}{5}\int_0^{12}(-128 + 19.2x)(16)dx + \tfrac{1}{2}\int_0^{16}(8x)(x)\,dx$$
$$= \frac{4096}{3} + 491.52 - 491.52 + \frac{16384}{3} = \frac{20480}{3}\text{ k-ft}^3\text{ (check)}$$

Note that the contribution of the bending moment on BC to the horizontal deflection of A is zero, as evidenced by the equal rotations of joints B and C in the revised elastic curve of Fig. 7.3f.

Example 7.2. Determine the horizontal and vertical deflections of the free end A of the cantilever rigid frame $ABCD$ shown in Fig. 7.4a, first by the moment-area method; then by the unit-load method.

Solution. The reactions at the fixed support (Fig. 7.4a) are computed to be

$$H_D = 8\text{ kips} \qquad V_D = 40\text{ kips} \qquad M_D = 40(12) - 8(8) = 416\text{ ft-kips}$$

The modified moment areas are computed from the moment diagram of Fig. 7.4b and shown in Fig. 7.4c; thus

$$A_1 = \tfrac{1}{3}(\tfrac{64}{3})(16) = \tfrac{1024}{9} \qquad A_2 = \tfrac{64}{5}(20) = 256$$
$$A_3 = \tfrac{1}{2}(\tfrac{480}{5})(12) = 576 \qquad A_4 = \tfrac{1}{2}(\tfrac{544}{2})(16) = 2176$$
$$A_5 = \tfrac{1}{2}(\tfrac{416}{2})(16) = 1664$$

Applying the moment-area theorems to the elastic curve DC' in Fig. 7.4d,

$$EI_c\theta_C = A_4 + A_5 = 2176 + 1664 = 3840$$
$$CC' = \frac{1}{EI_c}[A_4(\tfrac{16}{3}) + A_5(\tfrac{32}{3})] = \frac{88064}{3EI_c} = BB_1$$

From the geometry of the elastic curve $C'B'$,

$$EI_c\theta_B = EI_c\theta_C + A_2 + A_3 = 3840 + 256 + 576 = 4672$$
$$B_1B' = 20\theta_C + \frac{A_2}{EI_c}(10) + \frac{A_3}{EI_c}(16)$$
$$= \frac{1}{EI_c}[20(3840) + 256(10) + 576(16)] = \frac{88576}{EI_c} = A_1A'$$

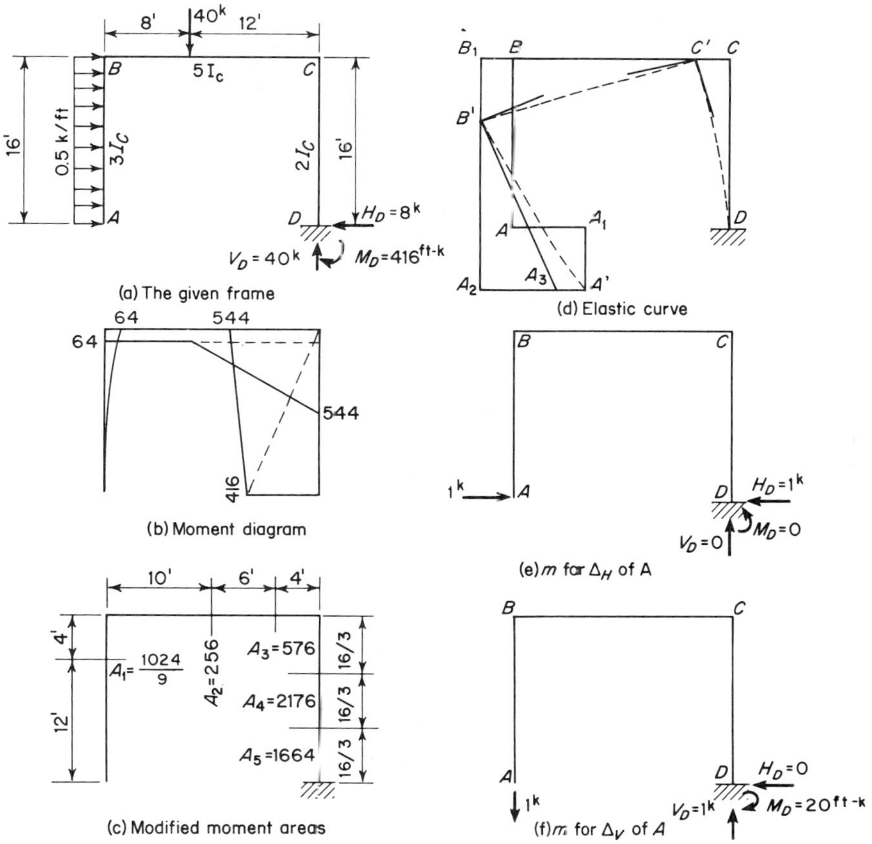

FIG. 7.4. The Rigid Frame of Example 7.2.

From the elastic curve $B'A'$,

$$AA_1 = A_2A_3 + A_3A' - BB_1 = 16\theta_B + \frac{A_1}{EI_c}(12) - \frac{88064}{3EI_c}$$

$$= \frac{1}{EI_c}\left[16(4672) + \frac{1024}{9}(12) - \frac{88064}{3}\right] = \frac{140288}{3EI_c}\ \text{k-ft}^3$$

When the unit-load method is used, the expressions for M, m for Δ_H of A, and m for Δ_V of A are taken from Fig. 7.4a, e and f.

$$EI_c\Delta_H \text{ of } A = \int \frac{Mm\,dx}{EI}$$

$$= \tfrac{1}{3}\int_0^{16}\left(-\frac{0.5x^2}{2}\right)(-x)\,dx + \tfrac{1}{5}\int_0^{8}(-64)(-16)\,dx$$

$$+ \tfrac{1}{5}\int_0^{12}(-64 - 40x)(-16)\,dx + \tfrac{1}{2}\int_0^{16}(416 + 8x)(x)\,dx$$

$$= \frac{4096}{3} + 1638.4 + 11673.6 + \frac{96256}{3} = \frac{140288}{3}\ \text{k-ft}^3 \text{ (check)}$$

$$
\begin{aligned}
EI_c\Delta_V \text{ of } A &= \int \frac{Mmdx}{EI} \\
&= \tfrac{1}{5}\int_0^8 (-64)(-x)\,dx + \tfrac{1}{5}\int_0^{12} (-64 - 40x)[-(x+8)]\,dx \\
&\quad + \tfrac{1}{2}\int_0^{16} (416 + 8x)(20)\,dx \\
&= 409.6 + (1228.8 + 5529.6 + 4608) + (66560 + 10240) \\
&= 88576 \text{ k-ft}^3 \text{ (check)}
\end{aligned}
$$

EXERCISE 7.1

Determine the horizontal deflection of the roller support at D of the statically determinate rigid frame $ABCD$ (Fig. 7.5), first by the moment-area conjugate-beam method; ~~then by the unit-load method.~~

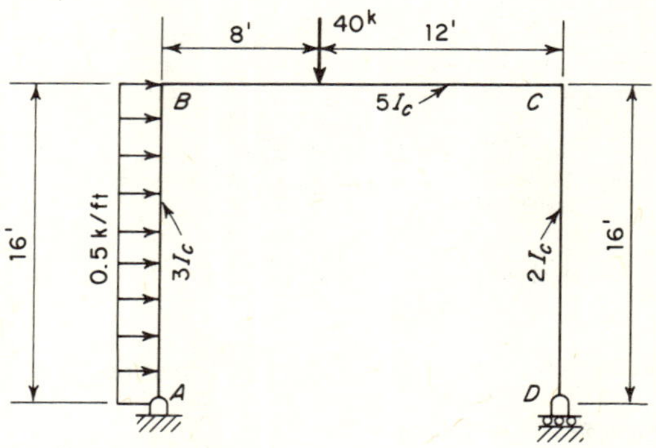

FIG. 7.5. Exercise 7.1.

EXERCISE 7.2

Determine the rotation, horizontal deflection, and vertical deflection of the free end D of the cantilever rigid frame $ABCD$ (Fig. 7.6), first by the moment-area method; then by the unit-load method.

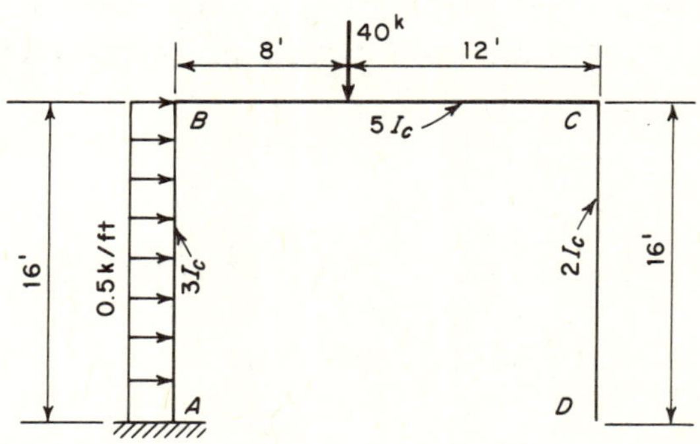

FIG. 7.6. Exercise 7.2.

7.3 Degree of Freedom vs. Degree of Indeterminacy

The degree of freedom of a rigid frame depends on the number of members and joints taken in the analysis procedure because the effect of adding a joint at any point on a straight member is simply to yield one additional member. For instance, even though the two rigid frames shown in Fig. 7.7a and b are exactly identical, one may be considered to have four joints and three members and the other, six joints and five members.

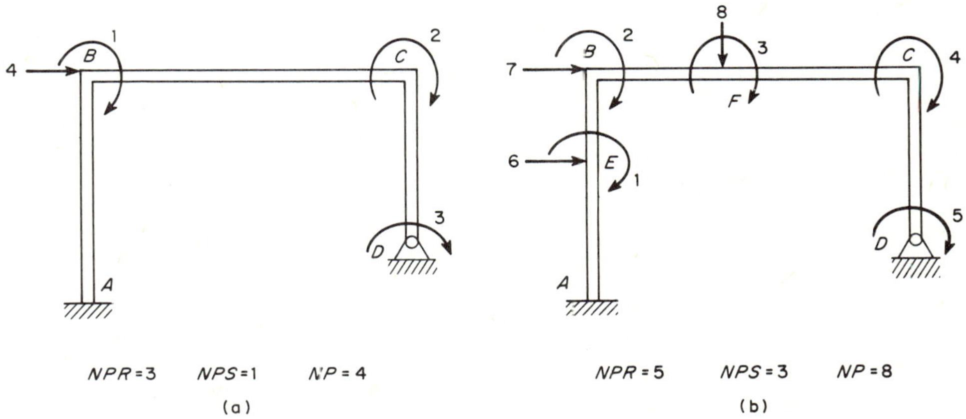

FIG. 7.7. Degree of Freedom of a Rigid Frame.

It is a simple matter to observe the number of unknown joint rotations *NPR*, but in assigning the number of independent unknown joint deflections *NPS*, it is important to keep in mind the two previously stated fundamental assumptions; namely, that axial deformation is ignored, and that transverse end displacements do not affect the member length. Thus, in Fig. 7.7a the only unknown joint deflection is the horizontal deflection of either joint *B* or *C*; in Fig. 7.7b the three unknown joint deflections are the horizontal deflection of *E*, the vertical deflection of *F*, and the horizontal deflection of any one of joints *B*, *F*, or *C*.

The degree of freedom *NP* of a rigid frame is dependent upon both the number of unknown joint rotations and the number of unknown independent joint deflections; or

$$NP = NPR + NPS \tag{7.3.1}$$

When axial deformation is neglected in the analysis, the axial forces in the members of a rigid frame are dependent on the member-end moments and can be computed from the laws of statics. This will require that just a sufficient number of resolution equations of equilibrium be available to determine these unknown axial forces. Let

$$NM = \text{number of members}$$
$$j = \text{number of joints including supports}$$
$$f = \text{number of fixed supports}$$
$$h = \text{number of hinged supports}$$
$$r = \text{number of roller supports}$$

The total number of resolution equations of equilibrium is $2j$; from this $(2f + 2h + r)$ conditions are needed for obtaining the linear-force reactions, and NPS conditions are required as equilibrium conditions in one-to-one correspondence with the unknown joint deflections; thus the remaining equations may be equated to the NM unknown axial forces, or

$$2j - (2f + h + r) - NPS = NM$$

from which

(7.3.2) $$NPS = 2j - (2f + 2h + r) - NM$$

It is generally desirable to make a graphical inspection of the number of independent joint deflections which are required to orient the deflected positions of all the joints and check this number against that yielded from Eq. (7.3.2).

Since now it has been ascertained that the only independent unknown internal forces are the two end moments acting on each member, the total number of internal forces becomes

(7.3.3) $$NF = 2 * NM$$

The degree of indeterminacy i is the excess of the number of internal forces over the number of equations of statics; or

(7.3.4) $$i = NF - NP = 2 * NM - (NPR + NPS)$$

Equation (7.3.4) is useful in verifying the number of redundants to be used in the force method of analysis.

7.4 Force Method—Reactions as Redundants

One common method of analyzing relatively simple statically indeterminate rigid frames is the application of the compatibility conditions to solve for the redundant reactions, equal in number to the degree of indeterminacy. For the rigid frame $ABCD$ shown in Fig. 7.8a,

$$i = 2 * NM - (NPR + NPS) = 6 - (4 + 1) = 1$$

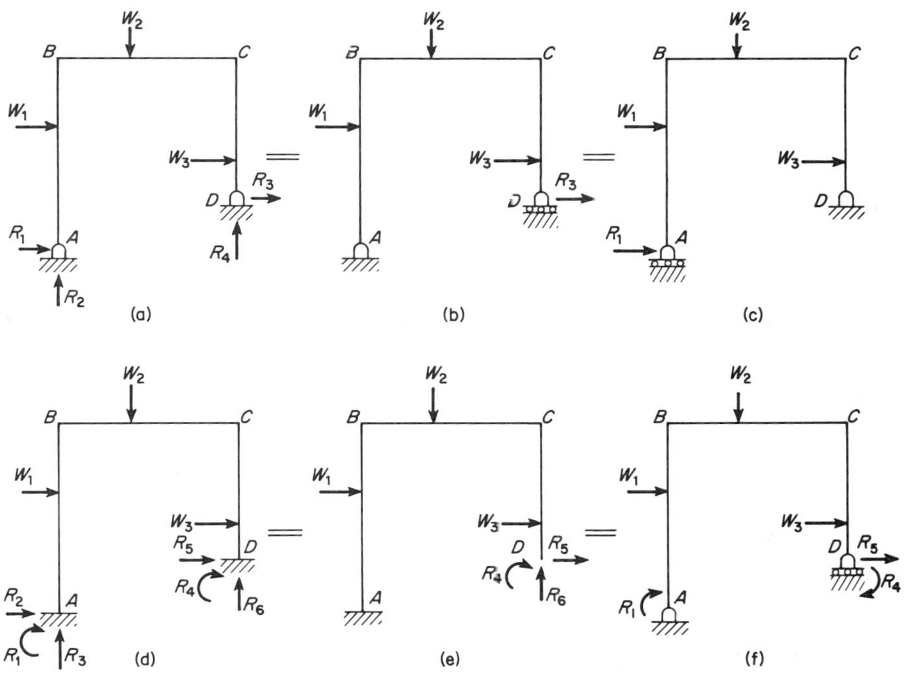

FIG. 7.8. The Force Method of Rigid Frame Analysis.

This rigid frame with two hinged supports may be considered to be equivalent to the degenerate frames of Fig. 7.8b or c, each subjected to the loads W_1-W_2-W_3 and the respective redundant reaction, R_3 or R_1. The redundant reaction (R_3 in Fig. 7.8b or R_1 in Fig. 7.8c) may be solved from the compatibility condition that the horizontal deflection of the roller support (at D in Fig. 7.8b or at A in Fig. 7.8c) must be zero.

Likewise, the three compatibility conditions in Fig. 7.8e are that R_4-R_5-R_6, together with the loads W_1-W_2-W_3, should make the rotation, horizontal deflection, and vertical deflection at D equal to zero; but in Fig. 7.8f, R_1-R_4-R_5, together with the loads W_1-W_2-W_3, should cause zero rotation at A, zero rotation at D, and zero horizontal deflection at D.

The *force method* of analyzing statically indeterminate rigid frames involves computation of joint rotations or deflections of the degenerate statically determinate rigid frames: first those due to the applied loads; then those due to each of the redundants. These joint displacements may be obtained by either the moment-area/conjugate-beam method, or the unit-load method, or both.

It should be realized that member-end moments may be used as the redundants instead of reactions. This approach could be useful in the advanced study of controlling the moments at some selected points to optimize the behavior of the whole structure. In this elementary text, it suffices to emphasize the concept of force method by its application to single-span, one-story rigid frames.

Example 7.3. Analyze the rigid frame of Fig. 7.9a by using the horizontal

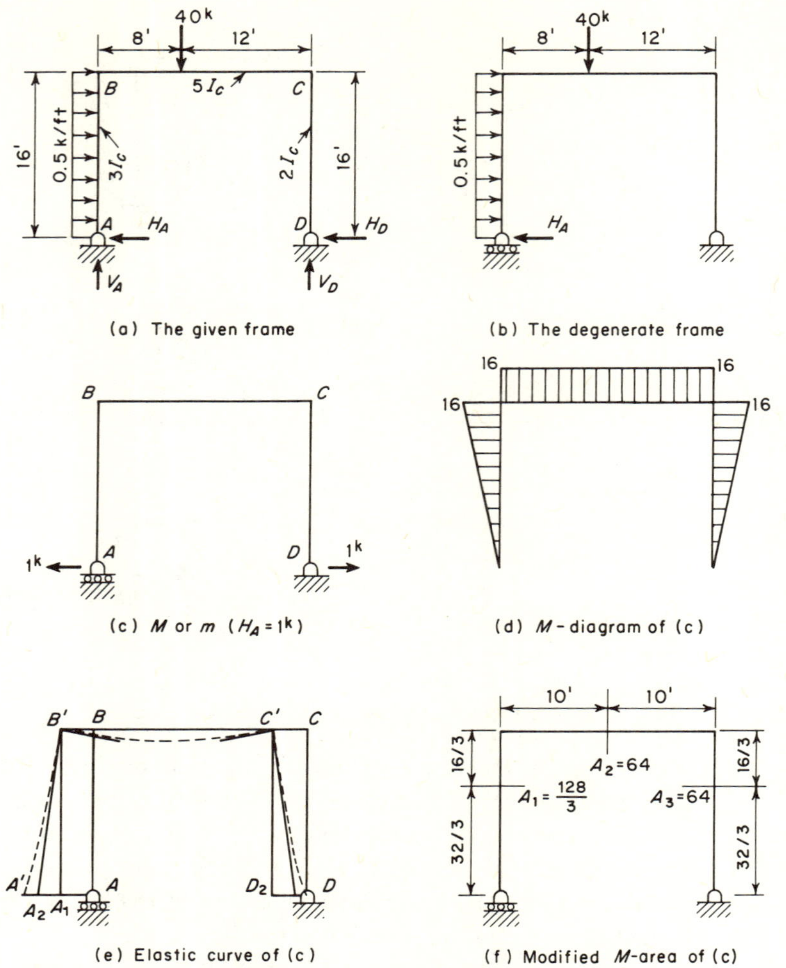

FIG. 7.9. The Force Method of Rigid Frame Analysis (Example 7.3).

reaction at A as the redundant. Draw shear and moment diagrams and compute the joint displacements in the final elastic curve.

Solution. The degenerate rigid frame of Fig. 7.9b has a roller support at A and is subjected to the applied loads and the horizontal force H_A. The horizontal deflection at A due to the applied loads may be quoted from the result of Example 7.1 to be

$$\Delta_H \text{ at } D \text{ due to loads} = \frac{20480}{3EI_c} \text{ k-ft}^3 \text{ (to the right)}$$

The horizontal deflection at A due to $H_A = 1$ kip may be computed by the moment-area/conjugate-beam method by referring to Fig. 7.9c, d, e and f; thus

$$EI_c\theta_B = EI_c\theta_C = \tfrac{1}{2}A_2 = 32$$

$$D_2D = D_2D_1 + D_1D = 16\theta_C + \frac{A_3}{EI_c}\left(\frac{32}{3}\right)$$

$$= \frac{1}{EI_c}\left[16(32) + 64\left(\frac{32}{3}\right)\right] = \frac{3584}{3EI_c} = CC' = BB'$$

$$AA' = AA_1 + A_1A_2 + A_2A' = BB' + 16\theta_B + \frac{A_1}{EI_c}\left(\frac{32}{3}\right)$$

$$= \frac{1}{EI_c}\left[\frac{3584}{3} + 16(32) + \frac{128}{3}\left(\frac{32}{3}\right)\right] = \frac{19456}{9EI_c}$$

Or, taking m from Fig. 7.9c and using the unit-load method,

$EI_c\Delta_H$ at A due to $H_A = 1$ kip

$$= \tfrac{1}{3}\int_0^{16} (x)^2\,dx + \tfrac{1}{5}\int_0^{20} (16)^2\,dx + \tfrac{1}{2}\int_0^{16} (-x)^2\,dx = \frac{19456}{9}\text{ k-ft}^3$$

Applying the compatibility condition that the horizontal deflection at A of the degenerate frame should be zero,

$$\frac{20480}{3} - \frac{19456}{9}H_A = 0$$

from which

$$H_A = \tfrac{60}{19} = 3.158 \text{ kips (to the left)}$$

Applying the laws of statics to the given frame of Fig. 7.9a,

$$H_D = 8 - 3.158 = 4.842 \text{ kips}$$

$$V_A = \frac{40(12) - 8(8)}{20} = 20.8 \text{ kips}$$

$$V_D = \frac{40(8) + 8(8)}{20} = 19.2 \text{ kips}$$

Having obtained the reactions, the analyst should draw the shear and moment diagrams and compute the joint displacements in the final elastic curve as exhibited in Fig. 7.10. The moment values at the critical locations shown in Fig. 7.10c are computed by the method of adding the shear areas shown in Fig. 7.10b.

Note how the elastic curve of Fig. 7.10d is sketched to match the moment diagram of Fig. 7.10c. Critical values of the elastic curve may be computed in the following order:

1) *Compute θ_B and θ_C by applying Conjugate-Beam Theorem I to $B'C'$;*

$$EI_c\theta_B = \tfrac{32}{60}A_1 - \tfrac{2}{3}A_2 - \tfrac{1}{3}A_3 = \tfrac{32}{60}(384) - \tfrac{2}{3}(26.94) - \tfrac{1}{3}(154.94)$$
$$= 135.19 \text{ k-ft}^2 \text{ (clockwise)}$$

$$EI_c\theta_C = \tfrac{28}{60}A_1 - \tfrac{1}{3}A_2 - \tfrac{2}{3}A_3 = \tfrac{28}{60}(384) - \tfrac{1}{3}(26.94) - \tfrac{2}{3}(154.94)$$
$$= 66.93 \text{ k-ft}^2 \text{ (counterclockwise)}$$

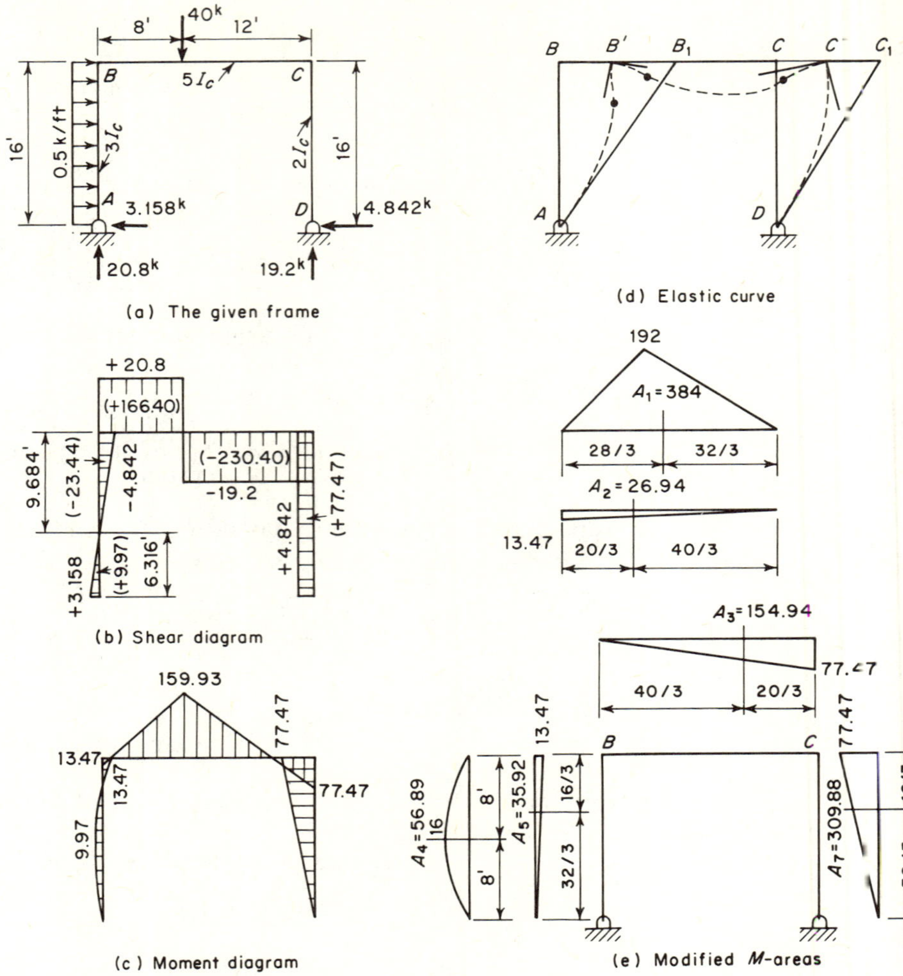

FIG. 7.10. Elastic Curve and Compatibility Check (Example 7.3).

2) *Compute θ_A and θ_D by applying Moment-Area Theorem I to B′A′ and D′C′, respectively*;

$$EI_c\theta_A = EI_c\theta_B + A_4 - A_5 = 135.19 + 56.89 - 35.92$$
$$= 156.16 \text{ k-ft}^2 \text{ (clockwise)}$$
$$EI_c\theta_D = A_6 - EI_c\theta_C = 309.88 - 66.93$$
$$= 242.95 \text{ k-ft}^2 \text{ (clockwise)}$$

3) *Compute BB′ and CC′ by applying Moment-Area Theorem II to A B′ and D′C′, respectively*;

$$BB' = BB_1 - B_1B' = 16\theta_A - \frac{A_4}{EI_c}(8) + \frac{A_5}{EI_c}\left(\frac{16}{3}\right)$$
$$= \frac{1}{EI_c}[16(156.16) - 56.89(8) + 35.92(16/3)] = \frac{2235.0}{EI_c}\text{k-ft}^3$$

$$CC' = CC_1 - C_1C' = 16\theta_D - \frac{A_6}{EI_c}\left(\frac{16}{3}\right)$$

$$= \frac{1}{EI_c}[16(242.95) - 309.88(16/3)] = \frac{2234.5}{EI_c}\text{k-ft}^3$$

The fact that BB' and CC' are equal, as computed independently above, assures that the compatibility condition for this singly-indeterminate structure has been satisfied.

Example 7.4. Analyze the rigid frame of Fig. 7.11a by using the horizontal and vertical reactions at A as the redundants. Draw shear and moment diagrams and compute the joint displacements in the final elastic curve.

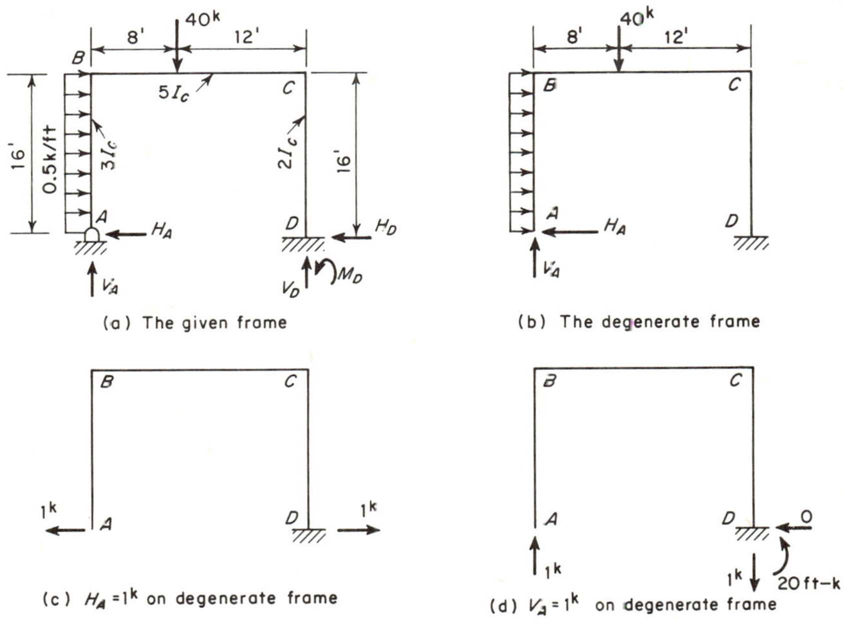

FIG. 7.11. The Force Method of Rigid Frame Analysis (Example 7.4).

Solution. The degenerate cantilever frame of Fig. 7.11b has a fixed support at D and is subjected to the applied loads and the two redundant forces H_A and V_A. The horizontal and vertical deflections at A due to the applied loads may be quoted from the results of Example 7.2:

$$\Delta_H \text{ at } A \text{ due to loads} = \frac{140288}{3EI_c}\text{ k-ft}^3 \text{ (to the right)}$$

$$\Delta_V \text{ at } A \text{ due to loads} = \frac{88576}{EI_c}\text{ k-ft}^3 \text{ (downward)}$$

Due to $H_A = 1$ kip acting on the degenerate frame,

$$\Delta_H \text{ at } A \text{ due to } (H_A = 1 \text{ kip}) = \int \frac{(M \text{ in Fig. 7.11c})(m \text{ in Fig. 7.11c})\, dx}{EI}$$

$$= \frac{1}{3EI_c}\int_0^{16} (x)^2\, dx + \frac{1}{5EI_c}\int_0^{20} (16)^2\, dx + \frac{1}{2EI_c}\int_0^{16} (-x)^2\, dx$$

$$= \frac{19456}{9EI_c} \text{ k-ft}^3 \text{ (to the left)}$$

$$\Delta_V \text{ at } A \text{ due to } (H_A = 1 \text{ kip}) = \int \frac{(M \text{ in Fig. 7.11c})(m \text{ in Fig. 7.11c})\, dx}{EI}$$

$$= \frac{1}{5EI_c}\int_0^{20} (16)(x)\, dx + \frac{1}{2EI_c}\int_0^{16} (-x)(-20)\, dx$$

$$= \frac{1920}{EI_c} \text{ k-ft}^3 \text{ (upward)}$$

Due to $V_A = 1$ kip acting on the degenerate frame,

$$\Delta_H \text{ at } A \text{ due to } (V_A = 1 \text{ kip}) = \int \frac{(M \text{ in Fig. 7.11d})(m \text{ in Fig. 7.11c})\, dx}{EI}$$

$$= \frac{1}{5EI_c}\int_0^{20} (x)(16)\, dx + \frac{1}{2EI_c}\int_0^{16} (-20)(-x)\, dx$$

$$= \frac{1920}{EI_c} \text{ k-ft}^3 \text{ (to the left)}$$

Note that Δ_V at A due to $H_A = 1$ kip is equal to Δ_H at A due to $V_A = 1$ kip since M and m have only been interchanged in the unit-load formula. A more general treatment of the Law of Reciprocal Deflections is demonstrated in Chapter 8.

$$\Delta_V \text{ at } A \text{ due to } (V_A = 1 \text{ kip}) = \int \frac{(M \text{ in Fig. 7.11d})(m \text{ in Fig. 7.11d})\, dx}{EI}$$

$$= \frac{1}{5EI_c}\int_0^{20} (x)^2\, dx + \frac{1}{2EI_c}\int_0^{16} (-20)^2\, dx$$

$$= \frac{11200}{3EI_c} \text{ k-ft}^3 \text{ (upward)}$$

Applying the two compatibility conditions that the horizontal and vertical deflections at A of the degenerate frame due to the applied loads and the redundant forces H_A and V_A should both be zero,

$$+\frac{140288}{3EI_c} - \frac{19456}{9EI_c}H_A - \frac{1920}{EI_c}V_A = 0$$

$$-\frac{88576}{EI_c} + \frac{1920}{EI_c}H_A + \frac{11200}{3EI_c}V_A = 0$$

Solving the above two equations,

$$H_A = 1.030 \text{ kips (to the left)}$$
$$V_A = 23.196 \text{ kips (upward)}$$

Applying the laws of statics to the given frame of Fig. 7.11a,

$$H_D = 8 - 1.030 = 6.970 \text{ kips (to the left)}$$
$$V_D = 40 - 23.196 = 16.804 \text{ kips (upward)}$$
$$M_D = 40(8) + 8(8) - 16.804(20) = 47.92 \text{ ft-kips (counterclockwise)}$$

The free-body, shear, and moment diagrams of the given frame are shown in Fig. 7.12a, b and c. Note that in the process of determining the critical values

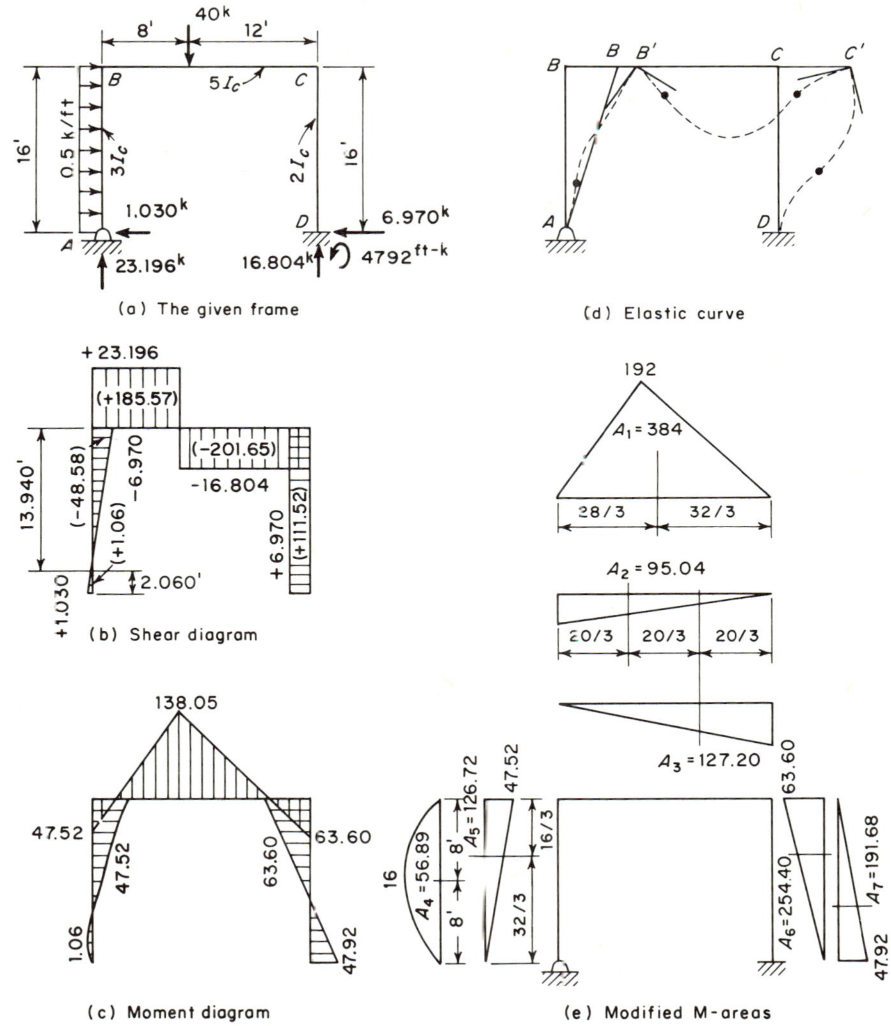

FIG. 7.12. Elastic Curve and Compatibility Check (Example 7.4).

on the moment diagram by the method of adding shear areas, a check is obtained in that the moment at D is in fact equal to 47.92 ft-kips causing compression on the right side of member DC.

Regarding the elastic curve shown in Fig. 7.12c, θ_B and θ_C are first computed by applying Conjugate-Beam Theorem I to $B'C'$; thus,

$$EI_c\theta_B = \tfrac{32}{60}A_1 - \tfrac{2}{3}A_2 - \tfrac{1}{3}A_3 = \tfrac{32}{60}(384) - \tfrac{2}{3}(95.04) - \tfrac{1}{3}(127.20)$$
$$= 99.04 \text{ k-ft}^2 \text{ (clockwise)}$$
$$EI_c\theta_C = \tfrac{28}{60}A_1 - \tfrac{1}{3}A_2 - \tfrac{2}{3}A_3 = \tfrac{28}{60}(384) - \tfrac{1}{3}(95.04) - \tfrac{2}{3}(127.20)$$
$$= 62.72 \text{ k-ft}^2 \text{ (counterclockwise)}$$

Applying the two moment-area theorems to $B'A'$,

$$EI_c\theta_A = EI_c\theta_B + A_4 - A_5 = 99.04 + 56.89 - 126.72$$
$$= 29.21 \text{ k-ft}^2 \text{ (clockwise)}$$
$$EI_c(BB') = EI_c(BB_1 - B_1B') = 16(EI_c\theta_A) - 8A_4 + \tfrac{16}{3}A_5$$
$$= 16(29.21) - 8(56.89) + \tfrac{16}{3}(126.72)$$
$$= 688.08 \text{ k-ft}^3 \text{ (to the right)}$$

Applying the two moment-area theorems from the fixed support at D toward C',

$$EI_c\theta_C = A_6 - A_7 = 254.40 - 191.68$$
$$= 62.72 \text{ k-ft}^2 \text{ (counterclockwise)}$$
$$EI_c(CC') = \tfrac{32}{3}A_7 - \tfrac{16}{3}A_5 = \tfrac{32}{3}(191.68) - \tfrac{16}{3}(254.40)$$
$$= 687.79 \text{ k-ft}^3 \text{ (to the right)}$$

The two compatibility checks obtained here are: 1) θ_C computed from $B'C'$ is equal to θ_C computed from DC'; 2) BB' computed from AB' is equal to CC' computed from DC'. Note that the number of compatibility checks should be equal to the degree of indeterminacy.

EXERCISE 7.3

Analyze the rigid frame with two hinged supports (Fig. 7.13) by using the horizontal reaction at D as the redundant. Refer to Exercise 7.1 for the hori-

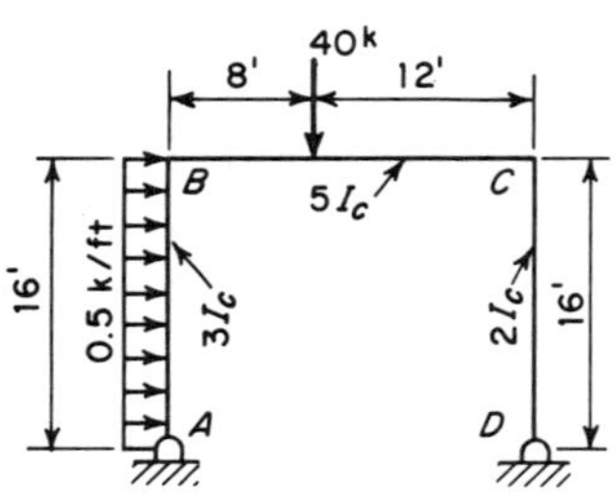

FIG. 7.13. Exercise 7.3.

zontal deflection at D of the degenerate structure due to the applied loads. Refer to Example 7.3 for the moment diagram and the elastic curve.

EXERCISE 7.4

Analyze the rigid frame with one fixed support and one hinged support (Fig. 7.14) by using the horizontal reaction at A and the moment reaction at D as the redundants. Refer to Example 7.1 for the joint rotations and deflections of the degenerate structure due to the applied loads. Refer to Example 7.4 for the moment diagram and the elastic curve.

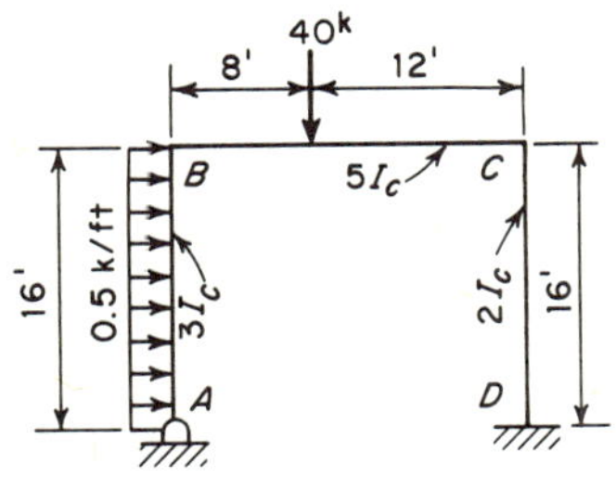

FIG. 7.14. Exercise 7.4.

EXERCISE 7.5

Analyze the rigid frame with two fixed supports (Fig. 7.15) by using the horizontal reaction, the vertical reaction, and the moment reaction at D as the redundants. Draw shear and moment diagrams and compute the joint displacements in the final elastic curve.

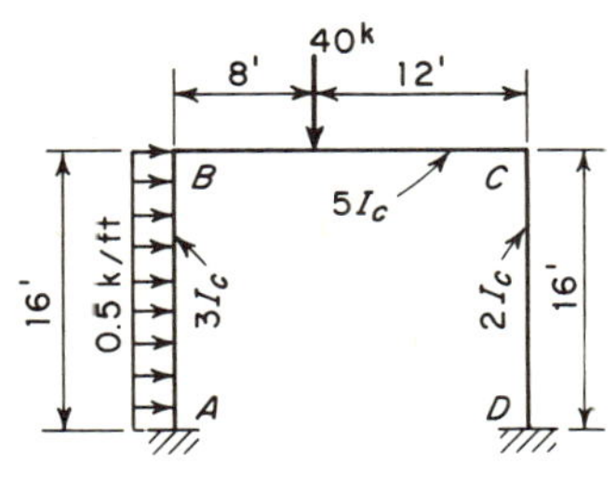

FIG. 7.15. Exercise 7.5.

7.5 Slope-Deflection Method vs. Matrix-Displacement Method

The slope-deflection method and the matrix-displacement method are almost identical; the only question of difference between the two lies in whether the algebraic equations are cast in conventional or matrix form. To demonstrate the equivalence of the two methods, it is desirable to adjust the notations in the slope-deflection equations (6.6.6a and b), as shown below, to those of the matrix-displacement method.

$$M_A = M_{0A} + \frac{4EI}{L}(\theta_A - R) + \frac{2EI}{L}(\theta_B - R)$$

$$M_B = M_{0B} + \frac{2EI}{L}(\theta_A - R) + \frac{4EI}{L}(\theta_B - R)$$

Referring to Fig. 7.16, the slope-deflection equations may be expressed as

(7.5.1a) $$F_i^* = F_{0i} + \frac{4EI}{L}e_i + \frac{2EI}{L}e_j$$

(7.5.1b) $$F_j^* = F_{0j} + \frac{2EI}{L}e_i + \frac{4EI}{L}e_j$$

in which

(7.5.1c) $$e_i = X_A - \frac{\Delta}{L}$$

(7.5.1d) $$e_j = X_B - \frac{\Delta}{L}$$

In these equations the letters "i" and "j" refer to member ends, wherein member end "i" enters into joint A with rotation X_A, and member end "j" enters into joint B with rotation X_B.

Equations (7.5.1a and b) are in fact the matrix equation

$$\{F^*\}_{2\times 1} = \{F_0\}_{2\times 1} + [S]_{2\times 2}\{e\}_{2\times 1}$$

for member ij and Eqs. (7.5.1c and d) are in fact the matrix equation

$$\{e\}_{2\times 1} = [B]_{2\times 1}\{X\}_{3\times 1}$$

for member ij.

The equilibrium conditions, equal in number to the degree of freedom or the number of unknown joint rotations and deflections, are in fact, rows of the matrix equation $\{P\} = \{A\}\{F\}$. In the slope-deflection method, these equations are established in terms of the total end moments $\{F^*\}$; but in the matrix-displacement method, these same conditions are satisfied by equating the resistance $\{AF\}$ offered by the end moments $\{F\}$ to the active joint forces $\{P\}$ in the directions of the degree of freedom, whereas these active joint forces are opposite to the locking joint forces required to maintain $\{X\} = 0$ in the fixed condi-

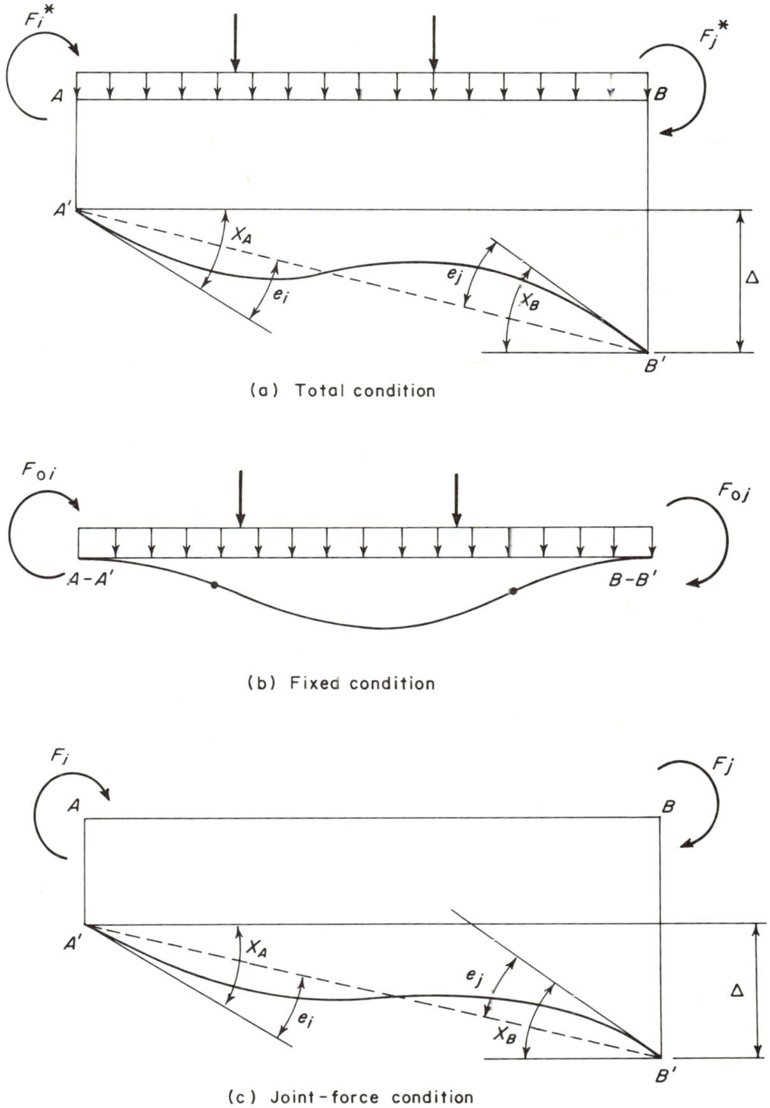

FIG. 7.16. Separation into Fixed and Joint-Force Conditions.

tion. Thus the matrix equation $\{P\} = [ASA^T]\{X\}$ contains the simultaneous equations in the slope-deflection method.

The matrix-displacement method is superior to the slope-deflection method in that the former divorces the applied loads from the establishment of the matrix $[ASA^T]$. When several loading conditions are to be analyzed, it is necessary only to expand the column matrix $\{P\}$ into a rectangular matrix. In addition, the matrix-displacement method can conveniently be written into a computer program, taking $[A]$, $[S]$, $[F_0]$, and $[P]$ as input, and giving $[X]$ and $[F^*]$ as output.

In the following example, the rigid frame, which has been solved by the force method in Example 7.3, will now be analyzed by the slope-deflection method and the matrix-displacement method. The reader will note that the two methods are identical in concept as well as in execution. The final statics and deformation checks are not made again because they have already been performed in the aforementioned example.

Example 7.5. Analyze the rigid frame with two hinged supports in Example 7.3, first by the slope-deflection method; then by the matrix-displacement method.

Solution. By considering the given frame of Fig. 7.17a to have three members and four joints, the degree of freedom in rotation is four, with θ_A-θ_B-θ_C-θ_D

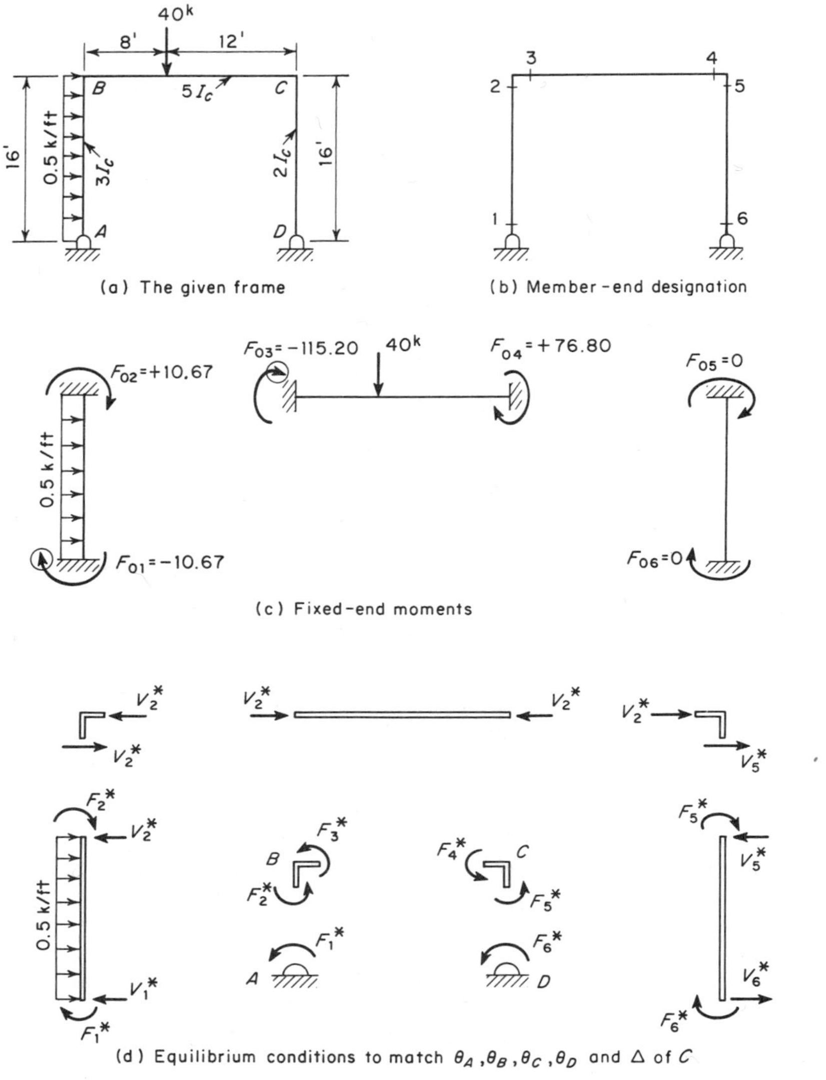

FIG. 7.17. The Slope-Deflection Method (Example 7.5).

as the unknown joint rotations; the degree of freedom in independent joint deflection is one, with the horizontal deflection Δ of joint C as the unknown deflection. One could have chosen the horizontal deflection Δ of joint B as the independent unknown deflection, but then the equilibrium condition would have been established by the horizontal resolution equation of equilibrium of joint B. Using Eq. (7.3.2)

$$NPS = 2j - (2f + 2h + r) - NM = 8 - (0 + 4 + 0) - 3 = 1$$

The slope-deflection equations are written according to the member-end designations of Fig. 7.17b and the fixed-end moments of Fig. 7.17c. In Fig. 7.17c, a circle is drawn around the clockwise arrowhead of the negative fixed-end moments to indicate that the real direction should be otherwise. The slope-deflection equations are

$$F_1^* = -10.67 + \frac{4E(3I_c)}{16}\left(\theta_A - \frac{\Delta}{16}\right) + \frac{2E(3I_c)}{16}\left(\theta_B - \frac{\Delta}{16}\right)$$

$$F_2^* = +10.67 + \frac{2E(3I_c)}{16}\left(\theta_A - \frac{\Delta}{16}\right) + \frac{4E(3I_c)}{16}\left(\theta_B - \frac{\Delta}{16}\right)$$

$$F_3^* = -115.20 + \frac{4E(5I_c)}{20}(\theta_B) - \frac{2E(5I_c)}{20}(\theta_C)$$

$$F_4^* = +76.80 + \frac{2E(5I_c)}{20}(\theta_B) + \frac{4E(5I_c)}{20}(\theta_C)$$

$$F_5^* = \quad 0. \quad + \frac{4E(2I_c)}{16}\left(\theta_C - \frac{\Delta}{16}\right) + \frac{2E(2I_c)}{16}\left(\theta_D - \frac{\Delta}{16}\right)$$

$$F_6^* \quad\quad 0. \quad + \frac{2E(2I_c)}{16}\left(\theta_C - \frac{\Delta}{16}\right) + \frac{4E(2I_c)}{16}\left(\theta_D - \frac{\Delta}{16}\right)$$

The equilibrium conditions, in one-to-one correspondence with the unknown joint rotations and deflection, are obtained by adding the *counter-clockwise* moments acting on joints A-B-C-D and the forces to the *left* acting on joint C and equating these summations to zero. Thus, referring to the free-body diagrams of the joints in Fig. 7.17d,

$$F_1^* = 0.$$

$$F_2^* + F_3^* = 0.$$

$$F_4^* + F_5^* = 0.$$

$$F_6^* = 0.$$

$$-V_2^* - V_5^* = -\left(4.0 + \frac{F_1^* + F_2^*}{16}\right) - \left(\frac{F_5^* + F_6^*}{16}\right) = 0$$

The following five simultaneous equations are obtained when the six slope-deflection equations are substituted into the five equilibrium conditions. It will be shown later that these equations are in fact identical to the matrix equation $[ASA^T]\{X\} = \{P\}$. Also, as previously proved in Art. 4.5, the matrix

	$EI_c\theta_A$	$EI_c\theta_B$	$EI_c\theta_C$	$EI_c\theta_D$	$EI_c\Delta$	
1	+3/4	+3/8			−9/128	= +10.67
2	+3/8	+7/4	+1/2		−9/128	= +104.53
3		+1/2	+3/2	+1/4	−3/64	= −76.80
4			+1/4	+1/2	−3/64	= 0.00
5	−9/128	−9/128	−3/64	−3/64	+15/1024	= +4.00

$[K] = [ASA^T]$ is the stiffness matrix of the whole structure and it is always symmetric. Solving the five simultaneous equations,

$$\begin{aligned} EI_c\theta_A &= +156.16 \text{ k-ft}^2 \\ EI_c\theta_B &= +135.19 \text{ k-ft}^2 \\ EI_c\theta_C &= -\ 66.93 \text{ k-ft}^2 \\ EI_c\theta_D &= +242.95 \text{ k-ft}^2 \\ EI_c\Delta &= +2234.8 \text{ k-ft}^3 \end{aligned}$$

Substituting these values of the joint displacements into the slope-deflection equations,

$$F_1^* = -10.67 + \tfrac{12}{16}(+156.16 - \tfrac{2234.8}{16}) + \tfrac{6}{16}(+135.19 - \tfrac{2234.8}{16}) = +0.01 \approx 0.$$
$$F_2^* = +10.67 + \tfrac{6}{16}(+156.16 - \tfrac{2234.8}{16}) + \tfrac{12}{16}(+135.19 - \tfrac{2234.8}{16}) = +13.48 \text{ ft-kips}$$
$$F_3^* = -115.20 + \tfrac{20}{20}(+135.19) + \tfrac{10}{20}(-66.93) = -13.47 \text{ ft-kips}$$
$$F_4^* = +76.80 + \tfrac{10}{20}(+135.19) + \tfrac{20}{10}(-66.93) = +77.47 \text{ ft-kips}$$
$$F_5^* = 0. + \tfrac{8}{16}(-66.93 - \tfrac{2234.8}{16}) + \tfrac{4}{16}(+242.95 - \tfrac{2234.8}{16}) = -77.48 \text{ ft-kips}$$
$$F_6^* = 0. + \tfrac{4}{16}(-66.93 - \tfrac{2234.8}{16}) + \tfrac{8}{16}(+242.95 - \tfrac{2234.8}{16}) = -0.01 \approx 0.$$

With these final end moments the free-body, shear and moment diagrams, as well as the critical values on the elastic curve, may be determined as shown in Fig. 7.10.

When the matrix-displacement method is used, the first task is the transfer of the transverse loads acting on the members to joint forces along the degrees of freedom. This is effected by using the fixed condition in which all joint displacements are zero, as shown in Fig. 7.18. At joint A, the action of the joint on the member to keep $X_1 = 0$ is 10.67 ft-kips counterclockwise; thus the active force P_1 on the joint is 10.67 ft-kips clockwise or $P_1 = +10.67$ ft-kips. At joint B, the action of the joint on the two member ends to keep $X_2 = 0$ is 115.20 ft-kips counterclockwise and 10.67 ft-kips clockwise, or a total of 104.53 ft-kips counterclockwise; thus, the active force P_2 on the joint is 104.53 ft-kips clockwise or $P_2 = +104.53$ ft-kips. Similarly $P_3 = -76.80$ ft-kips and $P_4 = 0$.

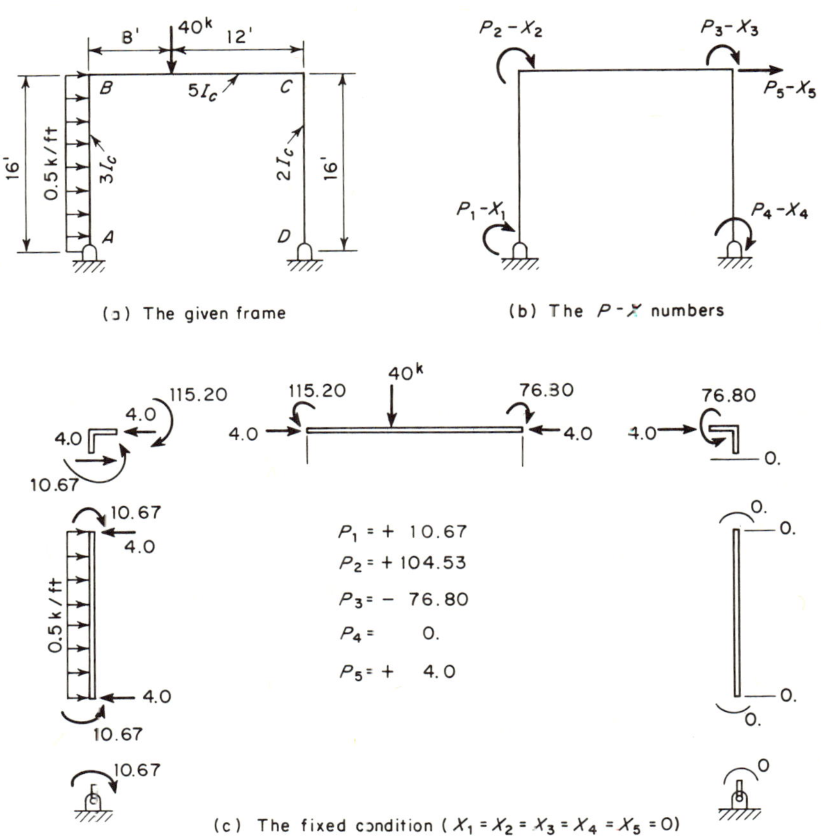

FIG. 7.18. The $\{P\}$ Matrix in Example 7.5.

The active force P_5 must be obtained at joint C since at the outset X_5 has been taken to be the horizontal deflection of joint C. In this context there should be horizontal equilibrium at joint B in the fixed condition. Thus in Fig. 7.18c, the summation of the active forces to the right at joint C is $P_5 = +4.0$ (from member CB) + 0.0 (from member CD) = +4.0 kips.

The statics matrix $[A]$ expresses the balance between the active joint forces $\{P\}$ and the resistance offered by the member-end moments $\{F\}$. Note that $\{F\}$ is different from $\{F^*\}$ from the relationship $\{F^*\} = \{F_0\} + \{F\}$. The five equations of statics, in one-to-one correspondence with the degrees of freedom, may be established by observing the free-body diagrams of the joints and members in Fig. 7.19, thus

$$P_1 = F_1$$

$$P_2 = F_2 + F_3$$

$$P_3 = F_4 + F_5$$

$$P_4 = F_6$$

$$P_5 = -\frac{F_1 + F_2}{16} - \frac{F_5 + F_6}{16}$$

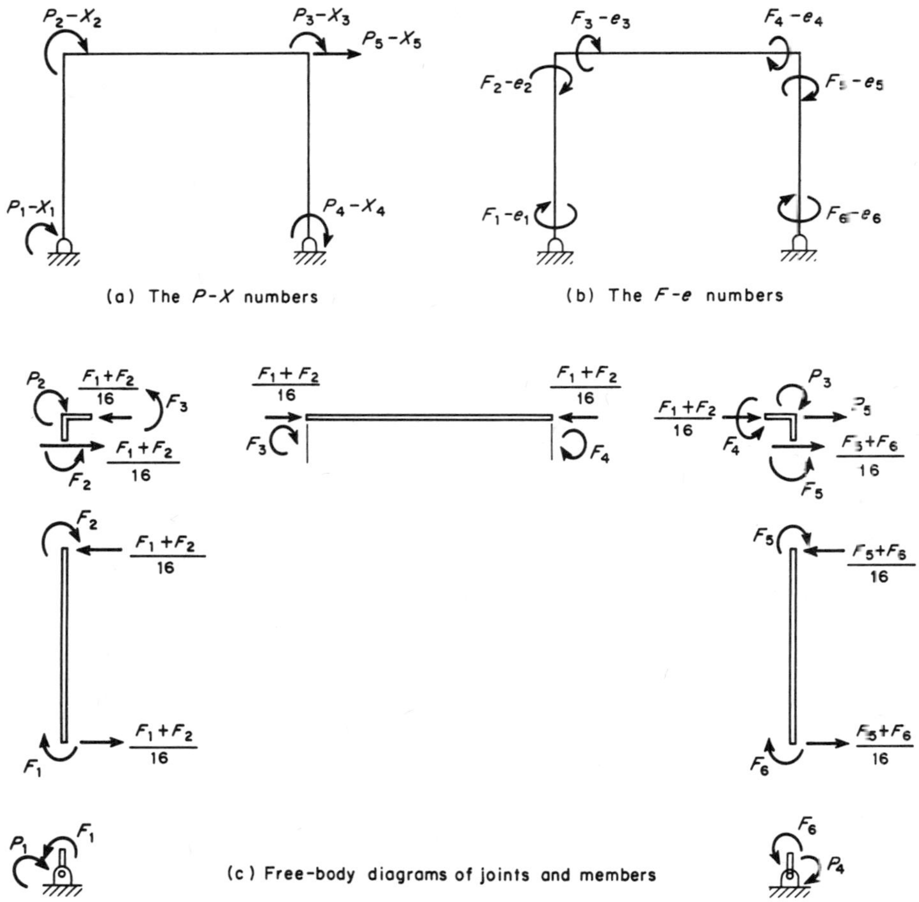

FIG. 7.19. The [A] Matrix in Example 7.5.

In matrix notation, the statics matrix [A] is

$[A]_{5\times 6} =$

P \ F	1	2	3	4	5	6
1	+1.0					
2		+1.0	+1.0			
3				+1.0	+1.0	
4						+1.0
5	$-\frac{1}{16}$	$-\frac{1}{16}$			$-\frac{1}{16}$	$-\frac{1}{16}$

The deformation matrix [B] expresses the member-end rotations (clockwise angle measured from the member axis to the tangent to the elastic curve) owing to each joint displacement. These effects may be conveniently assessed by observing the joint displacement diagrams in Fig. 7.20; thus

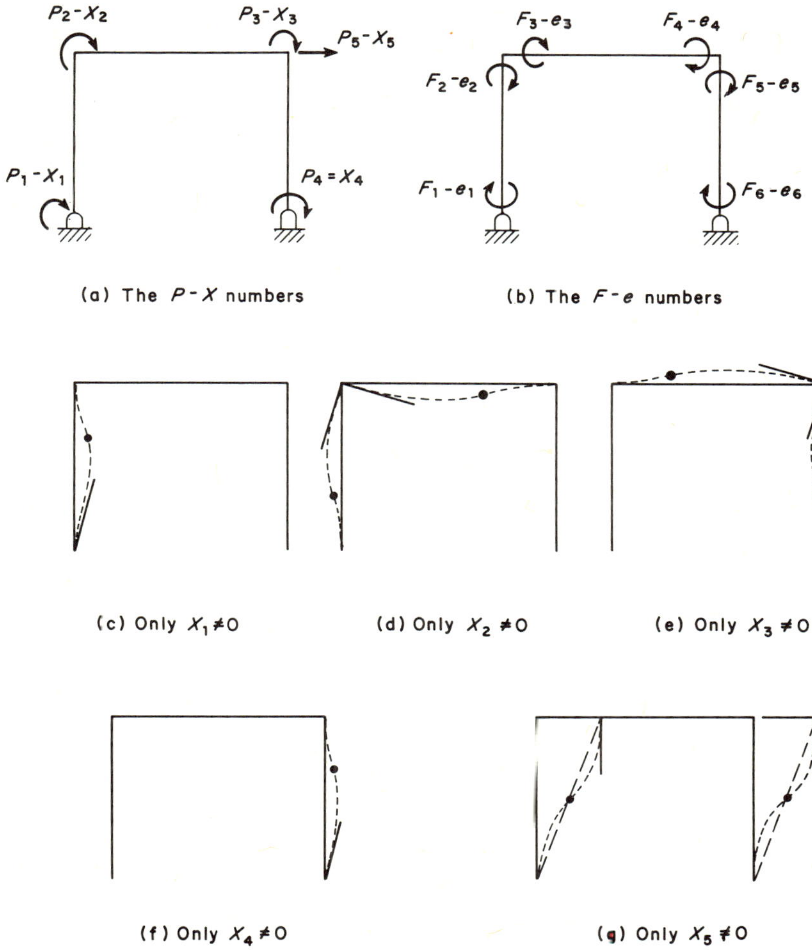

FIG. 7.20. The [*B*] Matrix in Example 7.5.

$[B]_{7\times 5} =$

e \ X	1	2	3	4	5
1	+1.0				$-\frac{1}{16}$
2		+1.0			$-\frac{1}{16}$
3		+1.0			
4			+1.0		
5			+1.0		$-\frac{1}{16}$
6				+1.0	$-\frac{1}{16}$

The [B] matrix is equal to the transpose of the [A] matrix, by the Principle of Virtual Work. Although not needed as an independent input matrix, it should be separately determined to check on the [A] matrix.

The member stiffness matrix [S] is, from Eq. (6.8.2),

$[S]_{6\times 6} = EI_c\,*$

F \ e	1	2	3	4	5	6
1	$\frac{3}{4}$	$\frac{3}{8}$				
2	$\frac{3}{8}$	$\frac{3}{4}$				
3			1	$\frac{1}{2}$		
4			$\frac{1}{2}$	1		
5					$\frac{1}{2}$	$\frac{1}{4}$
6					$\frac{1}{4}$	$\frac{1}{2}$

Applying matrix multiplication,

$[SA^T]_{6\times 5} = EI_c\,*$

F \ X	1	2	3	4	5
1	$+\frac{3}{4}$	$+\frac{3}{8}$			$-\frac{9}{128}$
2	$+\frac{3}{8}$	$+\frac{3}{4}$			$-\frac{9}{128}$
3		$+1$	$+\frac{1}{2}$		
4		$+\frac{1}{2}$	$+1$		
5			$+\frac{1}{2}$	$+\frac{1}{4}$	$-\frac{3}{64}$
6			$+\frac{1}{4}$	$+\frac{1}{2}$	$-\frac{3}{64}$

$[ASA^T]_{5\times 5} = EI_c\,*$

P \ X	1	2	3	4	5
1	$+\frac{3}{4}$	$+\frac{3}{8}$			$-\frac{9}{128}$
2	$+\frac{3}{8}$	$+\frac{7}{4}$	$+\frac{1}{2}$		$-\frac{9}{128}$
3		$+\frac{1}{2}$	$+\frac{3}{2}$	$+\frac{1}{4}$	$-\frac{3}{64}$
4			$+\frac{1}{4}$	$+\frac{1}{2}$	$-\frac{3}{64}$
5	$-\frac{9}{128}$	$-\frac{9}{128}$	$-\frac{3}{64}$	$-\frac{3}{64}$	$+\frac{15}{1024}$

Continuing with the typical procedure in the matrix displacement method,

$$X = [ASA^T]^{-1}P = \frac{1}{EI_c} * \begin{bmatrix} +156.16 \\ +135.19 \\ -\ \ 66.93 \\ +242.95 \\ +2234.80 \end{bmatrix}$$

and,

$$\{F^*\} = \{F_0\} + \{F\} = \{F_0\} + [SA^T]\{X\}$$

$$= \begin{bmatrix} -\ \ 10.67 \\ +\ \ 10.67 \\ -115.20 \\ +\ \ 76.80 \\ 0.00 \\ 0.00 \end{bmatrix} + \begin{bmatrix} +\ \ 10.67 \\ +\ \ \ 2.81 \\ +101.73 \\ +\ \ \ 0.67 \\ -\ \ 77.48 \\ 0.00 \end{bmatrix} = \begin{bmatrix} 0.00 \\ +\ \ 13.48 \\ -\ \ 13.47 \\ +\ \ 77.47 \\ -\ \ 77.48 \\ 0.00 \end{bmatrix}$$

The slight discrepancy between the pairs of numerical values of F_2^*-F_3^* or F_4^*-F_5^* is due to the fact that four or five significant figures have been used in the hand computations of this problem.

EXERCISE 7.6

Analyze the rigid frame of Exercise 7.4 (Fig. 7.14) with one fixed support and one hinged support first by the slope-deflection method; then by the matrix-displacement method.

EXERCISE 7.7

Analyze the rigid frame of Exercise 7.5 (Fig. 7.15) with two fixed supports first by the slope-deflection method; then by the matrix-displacement method.

7.6 Moment-Distribution Method

The moment-distribution method, as it has been applied to continuous beams, may be conveniently adapted to the analysis of rigid frames without unknown joint deflections. Had the two rigid frames shown in Fig. 7.21 been analyzed

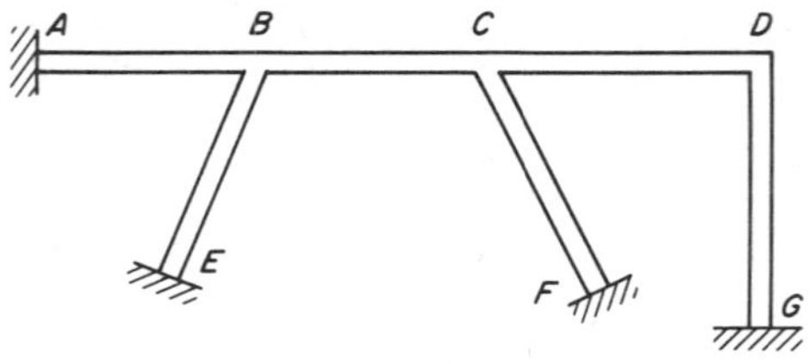

(a)

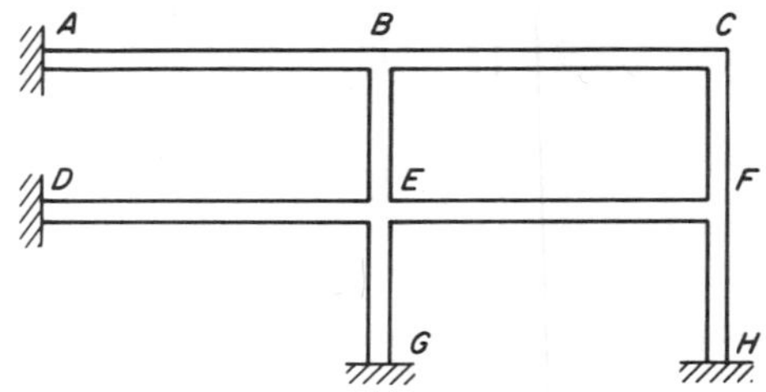

(b)

Joint	A	B			C			D		E	F	G
Member	AB	BA	BC	BE	CB	CD	CF	DC	DG	EB	FC	GD

(c) Heading for moment distribution of (a)

Joint	A	B			C		D	E				F			G	H
Member	AB	BA	BC	BE	CB	CF	DE	EB	ED	EF	EG	FC	FE	FH	GE	HF

(d) Heading for moment distribution of (b)

FIG. 7.21. Moment Distribution for Rigid Frames without Unknown Joint Deflections

by the slope-deflection or moment-distribution method, the only unknown joint displacements would have been those of rotations. Moment distribution allows the rotation of joints in graduated steps in order to balance the moments at the joints in the fixed condition. In setting up the moment distribution table, it is convenient to list the joints first, in alphabetical order, and then the members radiating from each joint, likewise in alphabetical order. Thus, the headings for the moment distribution of the rigid frames in Fig. 7.21a and b would be as shown in Fig. 7.21c and d. When an exterior joint is a hinged support, the method of modified stiffness factor and modified fixed-end moment discussed in Art. 6.13 applies as well. Likewise, the check on moment distribution discussed in Art. 6.12 should be made.

When the degree of freedom *NPS* in independent joint deflections is not zero, the moment distribution for the rigid frame may be performed in $(1 + NPS)$ stages. In the first stage, all degrees of freedom in joint deflection are restrained by providing fictitious supports such as those at B and E in Fig. 7.22b; the moment distribution is carried out, and the restraining forces R_{01} and R_{02} are obtained. In the second stage, one obtains a set of consistent fixed-end moments owing to an arbitrary joint deflection Δ_1, carries out the moment distribution, and computes the relieving forces P_{11} and P_{21}, as shown in Fig. 7.22c, from the balanced member-end moments. The third stage begins with a set of fixed-end moments consistent with an arbitrary joint deflection Δ_2 and the relieving forces P_{12} and P_{22}, as shown in Fig. 7.22d, are computed from the balanced member-end moments after moment distribution. Each required member-end moment in the given rigid frame of 7.22a is equal to the sum of its

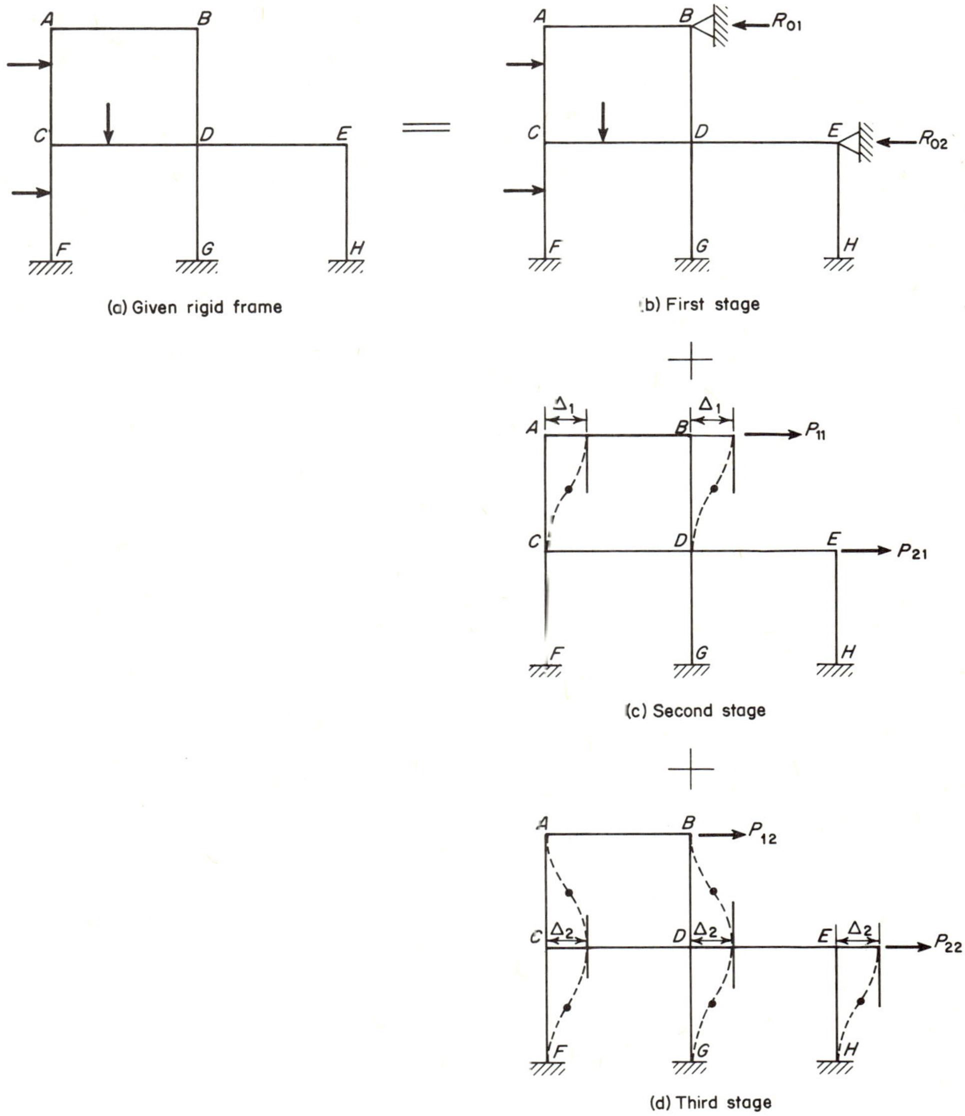

FIG. 7.22. Moment Distribution for Rigid Frames with Unknown Joint Deflections.

value in Fig. 7.22b, an unknown ratio k_1 times its value in Fig. 7.22c, and an unknown ratio k_2 times its value in Fig. 7.22d. The unknown ratios k_1 and k_2 may be determined from the conditions:

$$k_1 P_{11} + k_2 P_{12} = R_{01}$$
$$k_1 P_{21} + k_2 P_{22} = R_{02}$$

The actual amount of Δ_1 and Δ_2, if needed, may be obtained from the actual fixed-end moments in Fig. 7.22c and d, which are k_1 and k_2 times those fixed-end moments used at the beginning of moment distribution in the second and third stages, respectively. Likewise, each joint rotation in the given frame of Fig. 7.22a may be obtained as the sum of its value in Fig. 7.22b, k_1 times its value in Fig. 7.22c, and k_2 times its value in Fig. 7.22d, as these values are conveniently made available in the check procedure of each moment distribution table. Thus in the moment-distribution processes as outlined, not only the member-end moments, but also the joint displacements (including rotations and deflections) may be obtained. The entire moment distribution, then, is a relaxation procedure comprised of solving the simultaneous equations in the slope-deflection or matrix-displacement method.

Example 7.6. Analyze the rigid frame with two hinged supports in Example 7.3 or 7.5 by the moment-distribution method. Obtain the member-end moments as well as the joint rotations and deflections in the processes.

Solution. Because the degree of freedom in independent joint deflection of the given rigid frame shown in Fig. 7.23a is one, there is only need for one fictitious support in the first stage as shown in Fig. 7.23b. The first-stage moment distribution is shown in Table 7.1, wherein the method of modified stiffness factor and modified fixed-end moment is used. Note that an accuracy of two decimal places is decided upon at the outset in Table 7.1. The restraining force R_0 required at the fictitious support is computed as shown in Fig. 7.23c, by using the balanced moments in Table 7.1.

The absolute value of the fixed-end moments at the ends of a member subjected to a lateral deflection Δ, as shown in Fig. 7.23d, is $6EI\Delta/L^2$. Thus, the consistent fixed-end moments acting on the two columns to be used in the second stage should be in the ratio of $6E(3I_c)\Delta/16^2$ to $6E(2I_c)\Delta/16^2$, or of 3 to 2. Arbitrary values of -270 and -180, as shown in Fig. 7.23e, are used at the beginning of the second-stage moment distribution shown in Table 7.2. The relieving force P is computed as in Fig. 7.23f by using the balanced moments in Table 7.2.

The required ratio k is computed from the equation

$$R_0 = kP$$

or

$$k = \frac{R_0}{P} = \frac{6.142}{10.555} = 0.5819$$

The actual joint deflection Δ may be computed from the fixed-end moment formula referred to either the left or right column; thus

$$\frac{6E(3I_c)\Delta}{16^2} = 0.5819(270) \quad \text{or} \quad \frac{6E(2I_c)\Delta}{16^2} = 0.5819(180)$$

from which

$$\Delta = \frac{2234.5}{EI_c} \text{ k-ft}^3$$

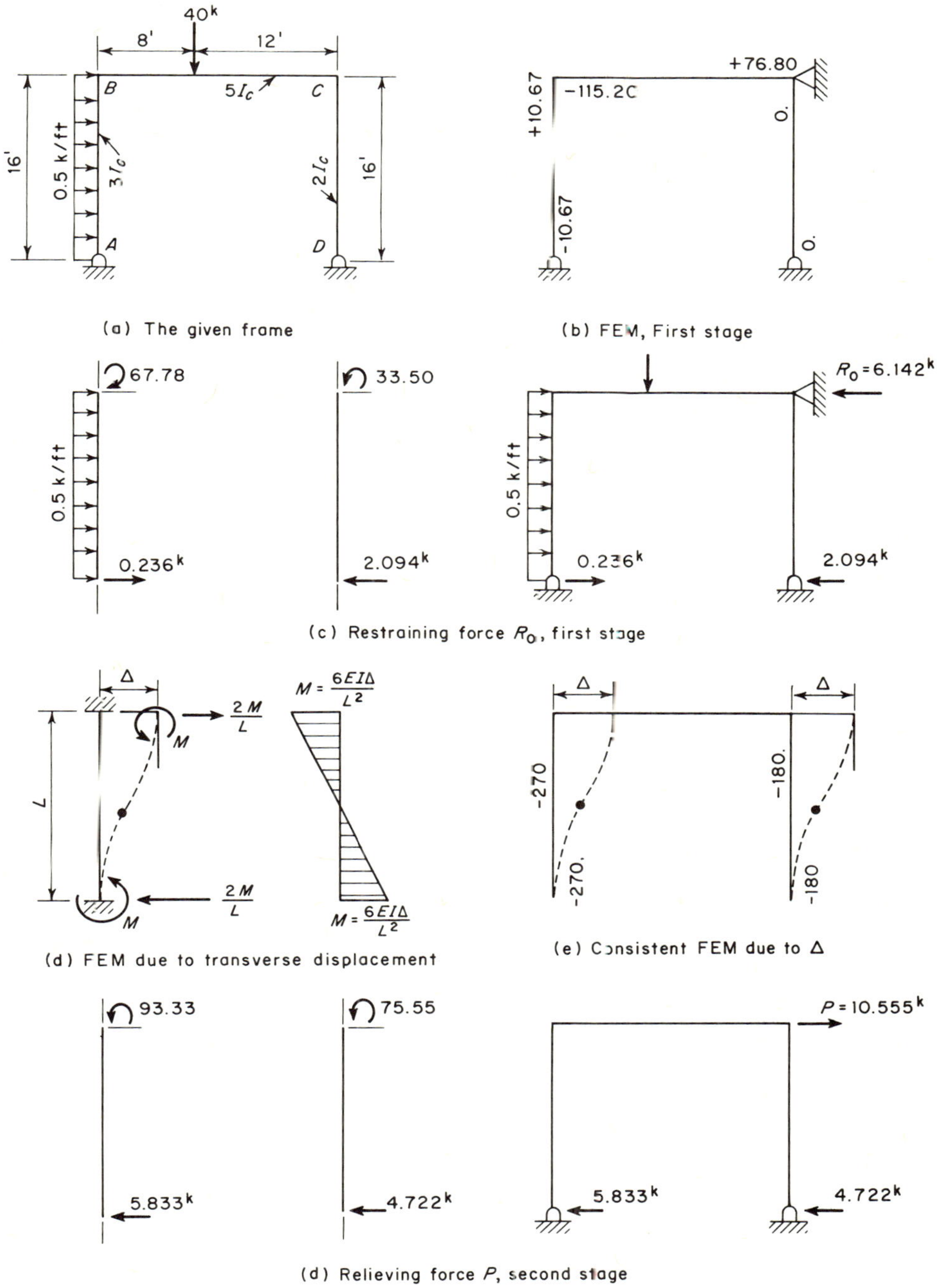

FIG. 7.23. The Moment-Distribution Method (Example 7.6).

The member-end moments for the given frame of Fig. 7.23a may now be obtained by summation of those of the first stage in Table 7.1 and 0.5819 times those of the second stage in Table 7.2. As shown in Table 7.3, these moments (0, +13.47, −13.47, +77.46, −77.46, 0) check well with the results of Example 7.3 or 7.5. Likewise, the joint rotations are the sums of those at the end of

TABLE 7.1 Moment Distribution, Example 7.6, First Stage

Joint	A	B		C		D
Member	AB	BA	BC	CB	CD	DC
$\frac{EI}{L}$	$\frac{3}{16}$	$\frac{3}{16}$	$\frac{5}{20}$	$\frac{5}{20}$	$\frac{2}{16}$	$\frac{2}{16}$
Rel. $\frac{EI}{L}$	3	3	4	4	2	2
Mod. Rel. $\frac{EI}{L}$	—	2.25	4	4	1.5	—
Mod. DF	—	0.3600	0.6400	0.7273	0.2727	—
FEM	−10.67	+10.67	−115.20	+76.80	—	—
Mod. FEM		+16.00	−115.20	+76.80	—	
Bal		+35.71	+ 63.49	−55.86	−20.94	
CO			− 27.93	+31.74		
Bal		+10.05	+ 17.88	−23.08	− 8.66	
CO			− 11.54	+ 8.94		
Bal		+ 4.15	+ 7.39	− 6.50	− 2.44	
CO			− 3.25	+ 3.70		
Bal		+ 1.17	+ 2.08	− 2.69	− 1.01	
CO			− 1.34	+ 1.04		
Bal		+ 0.48	+ 0.86	− 0.76	− 0.28	
CO			− 0.38	+ 0.43		
Bal		+ 0.14	+ 0.24	− 0.31	− 0.12	
CO			− 0.16	+ 0.12		
Bal		+ 0.06	+ 0.10	− 0.09	− 0.03	
CO			− 0.04	+ 0.05		
Bal		+ 0.01	+ 0.03	− 0.04	− 0.01	
CO			− 0.02	− 0.02		
Bal		+ 0.01	+ 0.01	+ 0.01	+ 0.01	
Balanced M	0.00	+67.78	− 67.78	+33.50	−33.50	0.00
Change	+10.67	+57.11	+ 47.42	−43.30	−33.50	0.00
$-\frac{1}{2}$ Change	−28.56	− 5.33	+ 21.65	−23.71	0.00	+16.75
Sum	−17.89	+51.78	+ 69.07	−67.01	−33.50	+16.75
$\frac{3EI}{L}$	$\frac{9}{16}$	$\frac{9}{16}$	$\frac{3}{4}$	$\frac{3}{4}$	$\frac{3}{8}$	$\frac{3}{8}$
$\theta = \frac{\text{Sum}}{3EI/L}$	−31.81	+92.05	+ 92.09	−89.35	−89.36	+44.66

TABLE 7.2 Moment Distribution, Example 7.6, Second Stage

Joint	A	B		C		D
Member	AB	BA	BC	CB	CD	DC
$\frac{EI}{L}$	$\frac{3}{16}$	$\frac{3}{16}$	$\frac{5}{20}$	$\frac{5}{20}$	$\frac{2}{16}$	$\frac{2}{16}$
Rel. $\frac{EI}{L}$	3	3	4	4	2	2
Mod. Rel. $\frac{EI}{L}$	—	2.25	4	4	1.5	—
Mod. DF	—	0.3600	0.6400	0.7273	0.2727	—
FEM	−270.00	−270.00	—	—	−180.00	−180.00
Mod. FEM Bal		−135.00 + 48.60	— +86.40	— +65.45	− 90.00 + 24.55	
CO Bal		 − 11.78	+32.72 −20.94	+43.20 −31.42	 − 11.78	
CO Bal		 + 5.66	−15.71 +10.05	−10.47 + 7.62	 + 2.85	
CO Bal		 − 1.37	+ 3.81 − 2.44	+ 5.02 − 3.65	 − 1.37	
CO Bal		 + 0.65	− 1.82 + 1.17	− 1.22 + .89	 + 0.33	
CO Bal		 − 0.16	+ .44 − .28	+ .58 − .42	 − 0.16	
CO Bal		 + 0.08	− .21 + .13	− .14 + .10	 + 0.04	
CO Bal		 − 0.02	+ .05 − .03	+ .06 − .04	 − 0.02	
CO Bal		 + 0.01	− .02 + .01	− .02 + .01	 + 0.01	
Balanced M	0.00	− 93.33	+93.33	+75.55	− 75.55	0.00
Change	+270.00	+176.67	+93.33	+75.55	+104.45	+180.00
$-\frac{1}{2}$ Change	− 88.34	−135.00	−37.78	−46.66	− 90.00	− 52.22
Sum	+181.66	+ 41.67	+55.55	−28.89	+ 14.45	+127.78
$\frac{3EI}{L}$	$\frac{9}{16}$	$\frac{9}{16}$	$\frac{3}{4}$	$\frac{3}{4}$	$\frac{3}{8}$	$\frac{3}{8}$
$\theta = \frac{\text{Sum}}{3EI/L}$	+322.9	+ 74.08	+74.07	+38.52	+ 38.53	+340.7

TABLE 7.3 Balanced Moments and Joint Rotations, Example 7.6

Joint	*A*	*B*		*C*		*D*
Member	*AB*	*BA*	*BC*	*CB*	*CD*	*DC*
Bal. *M* (1st stage) Bal. *M* (2nd stage)	0.00 0.00	+67.78 −54.31	−67.78 +54.31	+33.50 +43.96	−33.50 −43.96	0.00 0.00
Bal. *M*	0.00	+13.47	−13.47	+77.46	−77.46	0.00
θ (1st stage) θ (2nd stage)	−31.81 +187.90	+92.05 +43.11	+92.09 +43.10	−89.35 +22.41	−89.36 +22.42	+44.66 +198.25
θ	+156.11	+135.16	+135.19	−66.94	−66.94	+242.91

Table 7.1, and 0.5819 times those at the end of Table 7.2. As shown in Table 7.3, these values (+156.1, +135.2, −66.9, +242.9) also check well with the results of Example 7.3 or 7.5.

EXERCISE 7.8

Analyze the rigid frame with one fixed support and one hinged support of Exercise 7.4 (Fig. 7.14) by the moment-distribution method.

EXERCISE 7.9

Analyze the rigid frame with two fixed supports of Exercise 7.5 (Fig. 7.15) by the moment-distribution method.

7.7 Concluding Remarks

The treatment of statically indeterminate beams in the previous chapter has been extended to the analysis of statically indeterminate rigid frames. In the force method, the redundants have been chosen from the external reactions. On occasion, it may be desirable to use member-end moments as redundants, but this development will be deferred to more advanced study. In the displacement-method approach, the slope-deflection and matrix-displacement methods are essentially identical; in both cases the joint displacements, including rotations and deflections, are obtained by solving a system of linear equations. The moment distribution has again been shown to be a relaxation procedure of solving the linear equations in the displacement method.

Energy Theorems

8

8.1 Work-Energy vs. Physical-Geometric Methods

In first-order structural analysis involving only static loads in the elastic and linear range, one of two general approaches to the solution of a problem may be taken: the *work-energy method* or the *physical-geometric* method. In this text, the Principle of Virtual Work has been used to derive the unit-load method of finding the deflections of trusses, beams, and rigid frames in Arts. 3.6, 5.7, and 7.2. It has also been used to substantiate the transposition relationship between the statics and deformation matrices of trusses, beams, and rigid frames in Arts. 4.5, 6.7, and 7.5. On another occasion, in Art. 7.4, the Law of Reciprocal Deflections has been demonstrated by means of the unit-load method. Other than these brief instances of usages of or references to the work-energy concepts, the physical-geometric approach has been used in the major portion of this elementary textbook in relation to subjects as the joint-displacement equation for truss deflections, the moment-area method, force and displacement methods, as well as the slope-deflection and moment-distribution methods.

Insofar as only simple structures such as trusses, beams, and rigid frames are concerned, the physical-geometric approach may be relied upon exclusively for problem solution; the work-energy methods have been treated in parallel for elegance and refinement of the principles involved. However, when structural analysis is later extended to plate or shell structures, particularly in the recent and popular finite-element method, work-energy methods become the only approach which may be taken, since exact formulation of equilibrium and compatibility equations proves difficult if not impossible.

The purpose of this chapter is the gathering of the more important energy theorems in one place and the demonstration of their applications to trusses, beams, and rigid frames. Even though the same results could have been obtained by physical-geometric methods for these simple structures, one's confidence and acquaintance with the energy theorems might be established to such a degree *now* that future extension of their use in plate and shell structures would be facilitated.

8.2 The Principle of Real Work

The Principle of Real Work may be stated as: "The external work done on a linear elastic body must be equal to the internal work (sometimes also called strain energy, hence the name work-energy methods, or energy theorems) stored within the body."

If the linear elastic body is a two-dimensional truss as shown in Fig. 8.1,

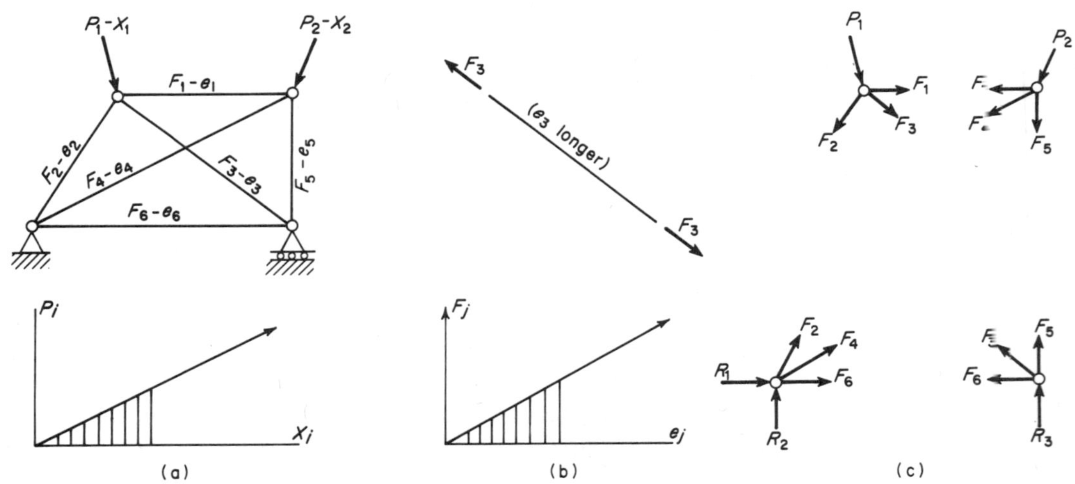

FIG. 8.1. External Work and Internal Work for a Truss.

the total external work done on it is $(\frac{1}{2}P_1X_1 + \frac{1}{2}P_2X_2)$ and the total internal work stored within it is $(\frac{1}{2}F_1e_1 + \frac{1}{2}F_2e_2 + \cdots + \frac{1}{2}F_6e_6)$; or in general

$$\Sigma\,(\tfrac{1}{2}P_iX_i) = \sum_{j=1}^{j=NF} (\tfrac{1}{2}F_je_j) \tag{8.2.1}$$

If the linear elastic body is a beam or a rigid frame as shown in Fig. 8.2, the total external work done on it is $(\frac{1}{2}P_1X_1 + \frac{1}{2}P_2X_2)$ and the total internal

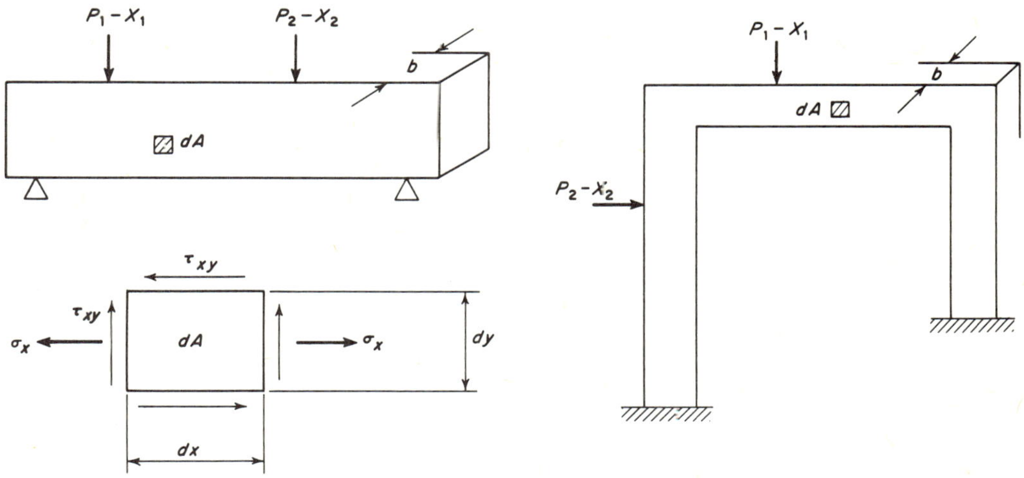

FIG. 8.2. External Work and Internal Work for a Beam or a Rigid Frame.

work stored within it is $\Sigma\,(\frac{1}{2}\sigma_x\epsilon_x + \frac{1}{2}\tau_{xy}\gamma_{xy})$ integrated over b times the entire two-dimensional area $dA = dydx$; or in general

$$\Sigma\,(\tfrac{1}{2}P_iX_i) = \int (\tfrac{1}{2}\sigma_x\epsilon_x + \tfrac{1}{2}\tau_{xy}\gamma_{xy})\,bdA \tag{8.2.2}$$

The contribution to the internal work by the shearing stresses τ_{xy} and strains γ_{xy}, and by the portion of the normal stresses σ_x and strains ϵ_x caused by the axial force on the bar, is small and for that reason has not been considered in this elementary textbook. Thus the right side of Eq. (8.2.2) becomes

$$\Sigma\,(\tfrac{1}{2}P_iX_i) = \int\!\!\int \frac{1}{2}\left(\frac{My}{I}\right)\left(\frac{My}{EI}\right) b\,dy\,dx = \int_0^L \frac{M^2\,dx}{2EI} \tag{8.2.3}$$

Little can be done towards effecting a general algebraic proof for the Principle of Real Work, other than stating that if the internal stresses and strains (e.g., F–e in Fig. 8.1 and σ_x-ϵ_x-τ_{xy}-γ_{xy} in Fig. 8.2) are in fact the "true" or "assumed" response due to a set of P–X application, the external work has to be equal to the internal work by the Law of Conservation of Work and Energy. In the case of trusses, however, so long as the first-order assumption that transverse end displacements do not change member length is asserted, the validity of Eq. (8.2.1) may be comprehended by seeing that the total work done by the zero resultant force acting on each of the four joints shown in Fig. 8.1c is zero; thus, noting that the work done by the forces F acting on the joint is always negative,

$$\begin{aligned}
&[\tfrac{1}{2}P_1X_1 - \tfrac{1}{2}(F_1)(\text{part of } e_1) - \tfrac{1}{2}(F_2)(\text{part of } e_2) - \tfrac{1}{2}(F_3)(\text{part of } e_3)] \\
&+ [\tfrac{1}{2}P_2X_2 - \tfrac{1}{2}(F_1)(\text{other part of } e_1) - \tfrac{1}{2}(F_4)(\text{part of } e_4) - \tfrac{1}{2}(F_5)(\text{part of } e_5)] \\
&+ [-\tfrac{1}{2}(F_2)(\text{other part of } e_2) - \tfrac{1}{2}(F_4)(\text{other part of } e_4) - \tfrac{1}{2}(F_6)(\text{part of } e_6)] \\
&+ [-\tfrac{1}{2}(F_3)(\text{other part of } e_3) - \tfrac{1}{2}(F_5)(\text{other part of } e_5) - \tfrac{1}{2}(F_6)(\text{other part of } e_6)] \\
&= 0
\end{aligned}$$

from which

$$\tfrac{1}{2}P_1X_1 + \tfrac{1}{2}P_2X_2 = \tfrac{1}{2}F_1e_1 + \tfrac{1}{2}F_2e_2 \cdots + \tfrac{1}{2}F_6e_6$$

This concept may be extended to Fig. 8.2 but one would have to stretch his imagination to deal with a two-dimensional continuum, rather than a finite number of joints and axially loaded bars.

If there is only one real load acting on a structure, the component of the deflection at the point of application and in the direction of that load can be determined by the Principle of Real Work.

Example 8.1. Determine, by the Principle of Real Work, the deflection at the free end of a prismatic cantilever beam subjected to a concentrated load at that end.

Solution. Using Eq. (8.2.3) and referring to Fig. 8.3,

$$\tfrac{1}{2}P\Delta = \int_0^L \frac{M^2\,dx}{2EI} = \int_0^L \frac{(-Px)^2\,dx}{2EI} = \frac{P^2L^3}{6EI}$$

from which

$$\Delta = \frac{PL^3}{3EI}$$

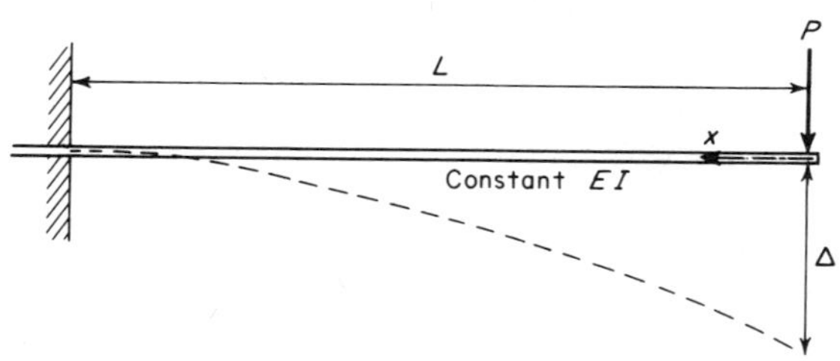

FIG. 8.3. Deflection of a Cantilever Beam.

Example 8.2. From the solution of the statically indeterminate truss in

Example 4.2, as shown completely in Fig. 4.6, compute and compare numerically the external work and the internal work.

Solution

$$\begin{aligned}
\text{External Work} &= [\tfrac{1}{2}(9)(41.08) + \tfrac{1}{2}(18)(109.03)] \times 10^{-3} \\
&= 1166.13 \times 10^{-3} \text{ in.-kips} \\
\text{Internal Work} &= \tfrac{1}{2}[7.73(7.73) + 11.88(11.88)(2) + 1.27(1.27) + 18(72) \\
&\quad + 1.69(6.76) + 2.11(4.22) + 4.81(9.62) \\
&\quad + 17.69(35.38)] \times 10^{-3} \\
&= 1166.05 \times 10^{-3} \text{ in.-kips} \quad \text{(check)}
\end{aligned}$$

EXERCISE 8.1

Determine, by the Principle of Real Work, the deflection under a concentrated load acting at the center of a prismatic simple beam.

EXERCISE 8.2

From the solution of the statically indeterminate truss in Exercise 4.1, compute and compare numerically the external work and the internal work.

8.3 The Principle of Virtual Work

The Principle of Virtual Work has been stated in Arts. 3.6, 5.7, and 7.2 as: The external work and the internal work, each done by the forces in the "equilibrium state" in going through the deformations of a "compatible state," must be equal.

An equilibrium state means that there is complete equilibrium of the structure in whole and in part. For instance, in the equilibrium state of Fig. 8.4a, it is defined that the two P's and the six F's are in equilibrium; how the F's might be determined is not the question. Similarly in Fig. 8.5a the normal stresses σ_x and the shear stresses τ_{xy} at all points have been defined and there is continuous equilibrium everywhere; again how values of σ_x and τ_{xy} might be obtained is not the matter in question.

A compatible state means that there is complete compatibility of deformation everywhere within the structure. For instance, in the compatible state of Fig. 8.4b, displaced locations A'-B'-D' of the original positions A-B-D have been defined, resulting in displacement components X_1 and X_2 in the directions of P_1 and P_2, and elongations e_1 to e_6 in the bars. Similarly in Fig. 8.5b, the deformed shape of the structure has been defined, resulting in displacement components X_1 and X_2, and normal and shear strains ϵ_x and γ_{xy}, respectively.

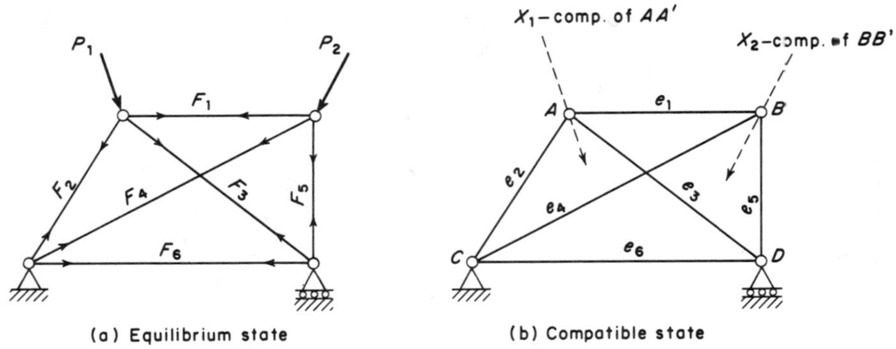

FIG. 8.4. The Equilibrium and Compatible States of a Truss.

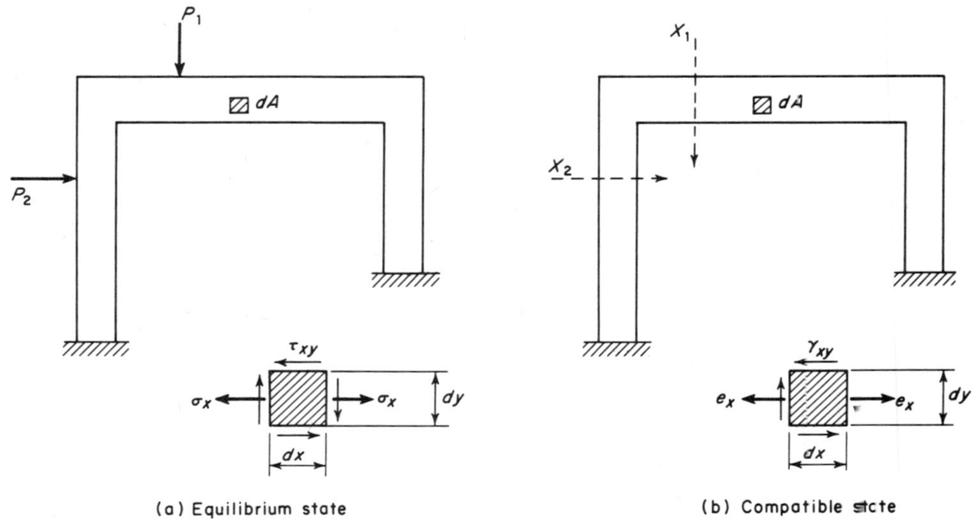

FIG. 8.5. The Equilibrium and Compatible States of a Rigid Frame.

The product of any one external force in the equilibrium state and the corresponding displacement component in the compatible state has similarity to the definition of work but it is not the actual work performed, thus the name *virtual work*. The sum of these products including all forces in the equilibrium state is defined as the *total external virtual work*, or more simply referred to as *external work*. Likewise, the summation, either of the products of the finite number of internal forces in the equilibrium state and their associated deformations in the compatible state (as in Fig. 8.4) or of the integration of the continuous product functions of internal stresses and strains (as in Fig. 8.5), is defined as the *total internal virtual work*, or more simply named, the *internal work*.

That the Principle of Virtual work is true may be explained by the following reasoning: Let the external forces P_1 and P_2 be first applied to the structure, e.g., as those in Figs. 8.4a or 8.5a. Then let the compatible state of Figs. 8.4b or 8.5b be superimposed on the deformed structure of Figs. 8.4a or 8.5a. So long as the material property is linear elastic and first-order analysis is assumed (the reader is urged to review Arts. 1.2 and 1.3), the forces and deformations in the resulting structure must be the sum of those in the separate states. Since the real external work and the real internal work must be equal not only in each of the two separate states but also in the combined state, the external work and internal work done by the forces in the equilibrium state in "riding" through the deformations in the compatible state during the superposition must be equal: thus, the Principle of Virtual Work.

If there is only one external force in the equilibrium state, the displacement component in the direction of that force in the compatible state can be obtained by applying the Principle of Virtual Work; hence, the unit-load method for trusses (Art. 3.6), beams (Art. 5.7), and rigid frames (Art. 7.2).

8.4 The Transposition Relationship between Statics and Deformation Matrices

In Arts. 4.5, 6.7, and 7.5, the statics and deformation matrices of trusses, beams, and rigid frames have been defined and the transposition relationship between them demonstrated. A unified proof for this relationship can be made by applying the Principle of Virtual Work to an equilibrium state described by the jth column of the statics matrix $[A]$ and a compatible state described by the ith column of the deformation matrix $[B]$ as follows:

$[A]_{NP\times NF} =$

$P \backslash F$	j
1	A_{1j}
2	A_{2j}
i	A_{ij}
NP	A_{NPj}

$[B]_{NF\times NP} =$

$e \backslash X$	i
1	B_{1i}
2	B_{2i}
j	B_{ji}
NP	B_{NPi}

Since X_i is the only nonzero external displacement in the compatible state, the total external virtual work is $(A_{ij}F_j)(X_i)$. Since F_j is the only nonzero internal force in the equilibrium state, the total internal virtual work is $(F_j)(B_{ji}X_i)$. Equating the external work and the internal work,

$$(A_{ij}F_j)(X_i) = (F_j)(B_{ji}X_i)$$

from which

$$A_{ij} = B_{ji}$$

Thus,

$$[A] = [B^T]$$

Referring to the truss of Fig. 8.6, the equality between the external virtual work and the internal virtual work can be expressed by

$$(A_{13}F_3)(X_1) = (F_3)(B_{31}X_1)$$

from which

$$A_{13} = B_{31}$$

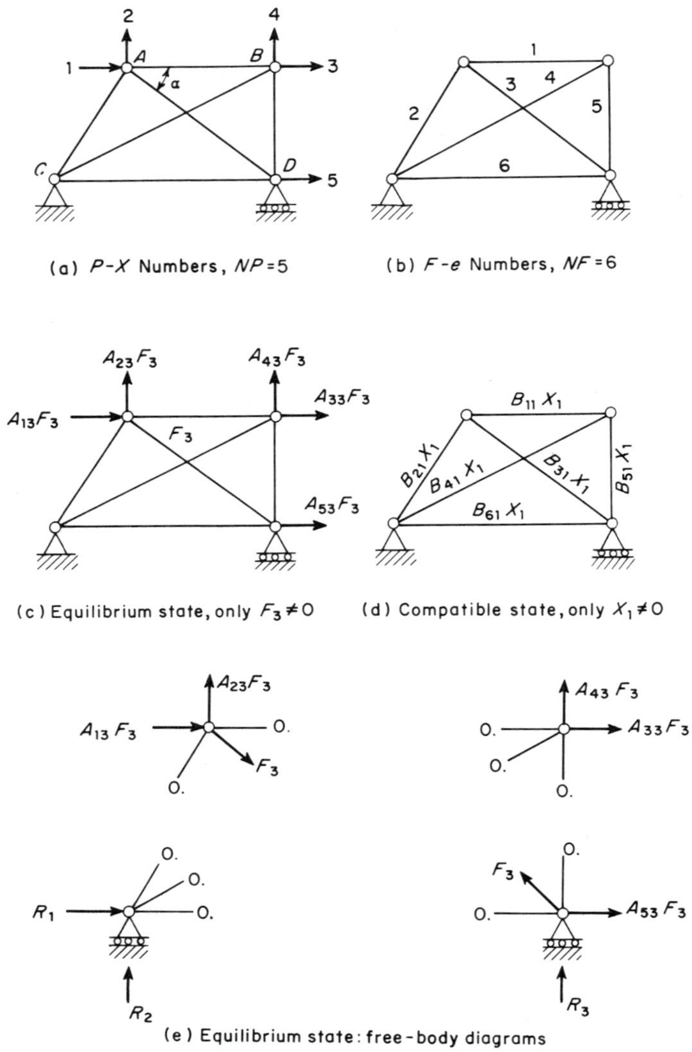

FIG. 8.6. The Principle of Virtual Work Applied to a Truss.

Both A_{13} and B_{31} can be independently observed to be $(-\cos\alpha)$. A specific proof can be made in this case by confirming that the work done by the zero resultant force acting on joint A (Fig. 8.6e) in going through the displacement X_1 must be zero.

Referring to the rigid frame of Fig. 8.7, the equality between the external virtual work and the internal virtual work can be expressed by

$$(A_{35}F_5)(X_3) = (F_5)(B_{53}X_3)$$

Each of A_{35} and B_{53} can be independently determined to be $(-1/L_{56})$. Here A_{35} is obtained by applying $\sum F_x = 0$ to joint B, but B_{53} is the counterclockwise rotation of the member axis, as this value is $(-1/L_{56})$ because the member axis

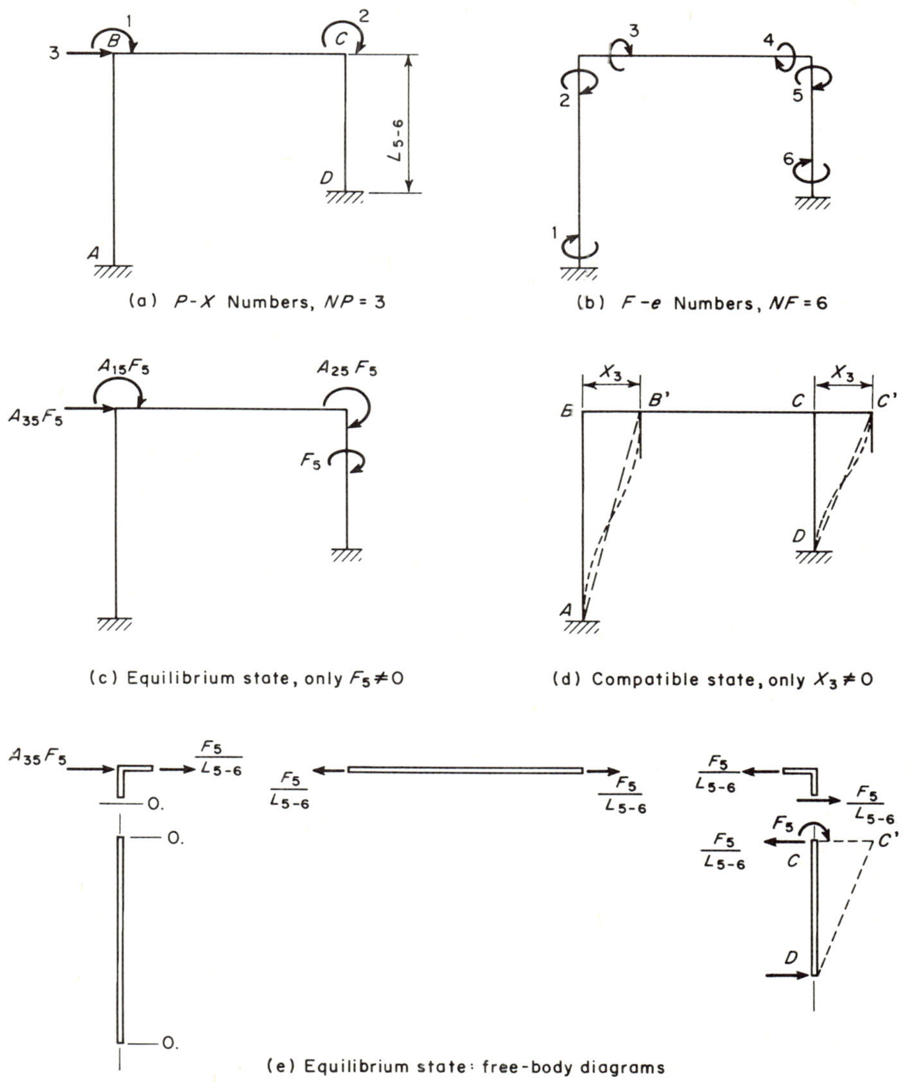

FIG. 8.7. The Principle of Virtual Work Applied to a Rigid Frame.

actually rotates in the clockwise direction due to X_3, following the equation

$$e_5 = X_2 - \frac{X_3}{L_{56}}$$

A specific proof for the virtual work equation in this case can be made by observing that the work done by the zero resultant force acting on member CD (Fig. 8.7e) in going through a rigid-body rotation must be zero.

8.5 The Reciprocal Energy Theorem

The Reciprocal Energy Theorem may be stated as: "The external or internal work, done by the external forces or internal stress resultants in a P-system, in going through the corresponding displacements or deformations in a Q-system, must be equal to the external or internal work done by the external forces or internal stress resultants in the Q-system, in going through the corresponding displacements or deformations in the P-system." In order that this theorem may be held to be true, the first-order method of analysis must be assumed and the material property must be linear elastic (see Arts. 1.2 and 1.3).

To prove the above theorem, one imagines that the P-system is first applied to the linear elastic body causing deformations (ΔP), and the resulting external or internal work may be symbolically represented by $\frac{1}{2}P(\Delta P)$. Then the Q-system is added, causing additional deformations (ΔQ); and the resulting external or internal work becomes

$$W = \tfrac{1}{2}P(\Delta P) + \tfrac{1}{2}Q(\Delta Q) + P(\Delta Q) \tag{8.5.1}$$

because the P-forces would have a "free ride" through the deformations (ΔQ). If, however, the Q-system had been applied first and then the P-system is added, the resulting external or internal work is

$$W = \tfrac{1}{2}Q(\Delta Q) + \tfrac{1}{2}P(\Delta P) + Q(\Delta P) \tag{8.5.2}$$

Since the total internal work within the linear elastic body under both the P- and Q-systems must be the same regardless of the order of application, equating Eq. (8.5.1) to Eq. (8.5.2) gives

$$P(\Delta Q) = Q(\Delta P) \tag{8.5.3}$$

Equation (8.5.3) is the Reciprocal Energy Theorem.

If there is only one single unit load in each of the P- and Q-systems and the notation δ_{PQ} is used to denote the deflection at P due to a unit load at Q, then by applying the Reciprocal Energy Theorem to each case of Fig. 8.8, one would obtain

$$(1.0)(\delta_{QP}) = (1.0)(\delta_{PQ})$$

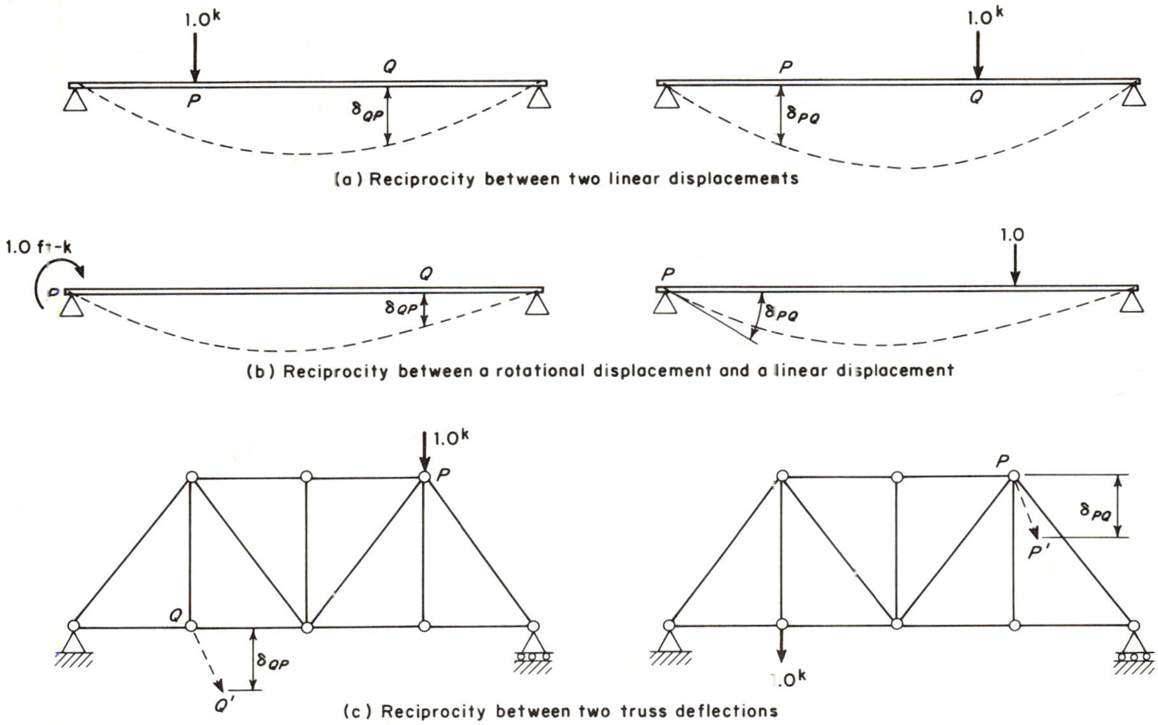

FIG. 8.8. The Law of Reciprocal Deflections.

or,

$$\delta_{QP} = \delta_{PQ} \tag{8.5.4}$$

Equation (8.5.4) is the Law of Reciprocal Deflections which states that the deflection at Q due to a unit load at P is equal to the deflection at P due to a unit load at Q. The reader is referred to Art. 7.4 where the Law of Reciprocal Deflections is demonstrated in the unit-load method of finding deflections; thus referring to Fig. 8.8a and b,

$$\delta_{PQ} = \int \frac{(m_Q)(m_P)\,dx}{EI}$$

$$\delta_{QP} = \int \frac{(m_P)(m_Q)\,dx}{EI}$$

and,

$$\delta_{QP} = \delta_{PQ}$$

In case of the truss as shown in Fig. 8.8c, by the unit-load method,

$$\delta_{PQ} = \sum F_P e_Q = \sum \frac{F_P F_Q L}{AE}$$

$$\delta_{QP} = \sum F_Q e_P = \sum \frac{F_Q F_P L}{AE}$$

and,

$$\delta_{QP} = \delta_{PQ}$$

Historically, the Law of Reciprocal Deflections is credited to Maxwell* in 1864 in his work on truss deflections by the unit-load method, but the more general Reciprocal Energy Theorem is due to the work of Betti** in 1872.

Example 8.3. Show that the Reciprocal Energy Theorem holds between the force systems of Fig. 4.6c and Fig. 4.6h and j.

Solution. The force systems of Fig. 4.6c and Fig. 4.6h and j are shown again in Fig. 8.9a and b and called the *P*- and *Q*-systems, respectively,

$$P * \Delta Q = [(9)(41.08) + (18)(109.03)] \times 10^{-3} = 2.332 \text{ in.-kips}$$

$$Q * \Delta P = [(9)(57.94) + (18)(139.33) - (12.46)(55.92)] = 2.322 \text{ in.-kips}$$

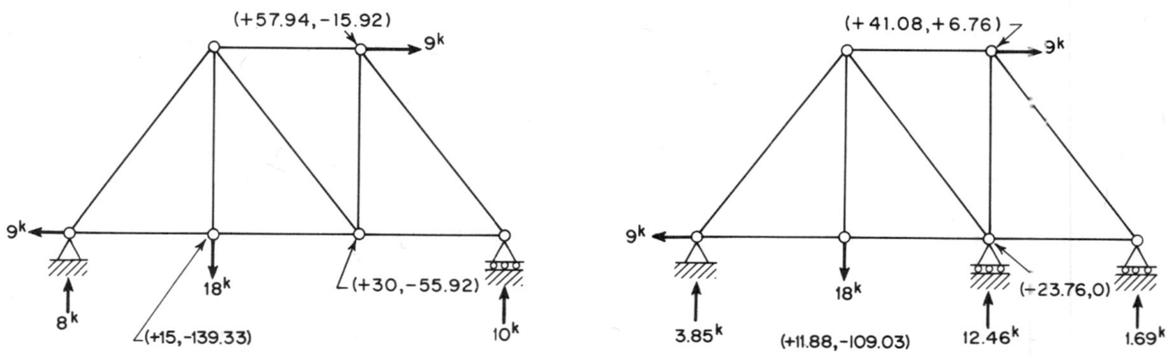

Numbers pointing at the joints are joint displacements in 10^{-3} in.

(a) *P*-System (b) *Q*-System

FIG. 8.9. The Reciprocal Energy Theorem Applied to a Truss.

8.6 Law of Reciprocal Forces vs. Law of Reciprocal Deflections

In Art. 4.5 it has been proved by means of matrix algebra that the external-force vs. joint-displacement matrix $[K] = [ASA^T]$ is a symmetric matrix. That K_{ij} and K_{ji} are equal may also be proved by applying the Reciprocal Energy

**History of Strength of Materials*, by S. P. Timoshenko, McGraw-Hill Book Company, 1953, p. 207.

**Ibid., p. 320.

Theorem to a P-system in which only $X_j \neq 0$ and a Q-system in which only $X_i \neq 0$ as shown below:

$[K] = [ASA^T] =$

P \ X	i	j
1		
i	K_{ii}	K_{ij}
j	K_{ji}	K_{jj}
NP		
	Q-system	P-system

Thus,

$$P * \Delta Q = Q * \Delta P$$

$$K_{ij}X_j * X_i = K_{ji}X_j * X_i$$

and,

$$K_{ij} = K_{ji} \tag{8.6.1}$$

Equation (8.6.1) may be called the Law of Reciprocal Forces which states that the force at the ith degree of freedom due to a unit jth displacement is equal to the force at the jth degree of freedom due to a unit ith displacement. Because of the fact that the $[K]$ matrix expresses external forces in terms of displacements, it is called the *structure stiffness matrix*.

As the inverse of a symmetric matrix is also symmetric, the matrix $[\delta] = [K^{-1}]$ is symmetric. That δ_{ij} and δ_{ji} are equal is indicated by the Law of Reciprocal Deflections (or Displacements). Because of the fact that the $[\delta] = [K^{-1}]$ matrix expresses displacements in terms of forces, it is called the *structure flexibility matrix*.

8.7 Extension to Nonlinear Material Property

Had the material property been nonlinear, the arguments presented in the preceding articles would have applied if one can imagine, first of all, a known parent state (whose external P-X and internal F-e values are known) represented by the big dot on the stress–strain curve of Fig. 8.10b. This instantaneous situation may be linearized provided that only compatible infinitesimal increments of external displacements ΔX and internal deformations Δe are added to this parent state.

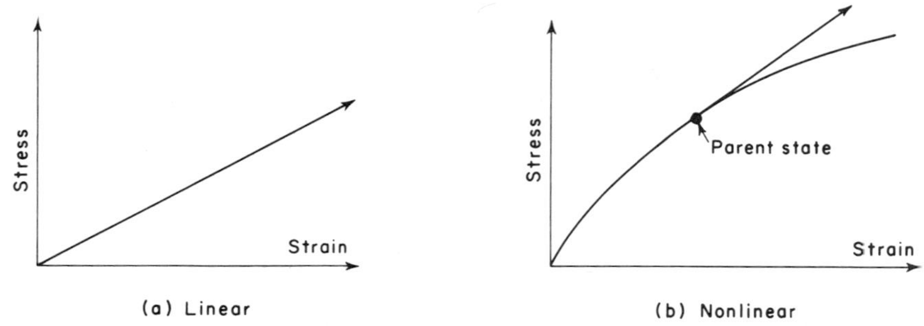

FIG. 8.10. Linear and Nonlinear Material Property.

The Principle of Virtual Work would become: "The incremental external and internal work, each done by the forces in the 'parent equilibrium state' in going through the infinitesimal deformations of an 'incremental compatible state,' must be equal." Or, in symbolic form

$$\Sigma P(\Delta x) = \Sigma F(\Delta e) \tag{8.7.1}$$

There would also be an *instantaneous structure stiffness matrix* $[K]$ at the parent state, which would express the incremental forces ΔP in terms of the incremental displacements ΔX. Likewise the *instantaneous structure flexibility matrix* $[\delta']$ would express the incremental displacements ΔX in terms of the incremental forces ΔP.

8.8 Potential and Complementary-Potential Energy vs. Strain and Complementary-Strain Energy

It has been stated in Art. 1.4 that any structure can be considered to be either an assemblage of springs (as in a truss) or a compaction of an infinite number of little springs (as in a beam). For the truss of Fig. 8.11a, which is composed of six *nonlinear* springs and subjected to two external forces, the definition of a parent state in which the sets of P–X and F–e values have been reached through the typical paths can be depicted by Fig. 8.11b and c. The total external work done on the truss is the sum of the two areas under the two P–X curves of Fig. 8.11b, and the total internal work is the sum of the six areas under the six F–e curves of Fig. 8.11c. Applying the Principle of Real Work to this nonlinear truss,

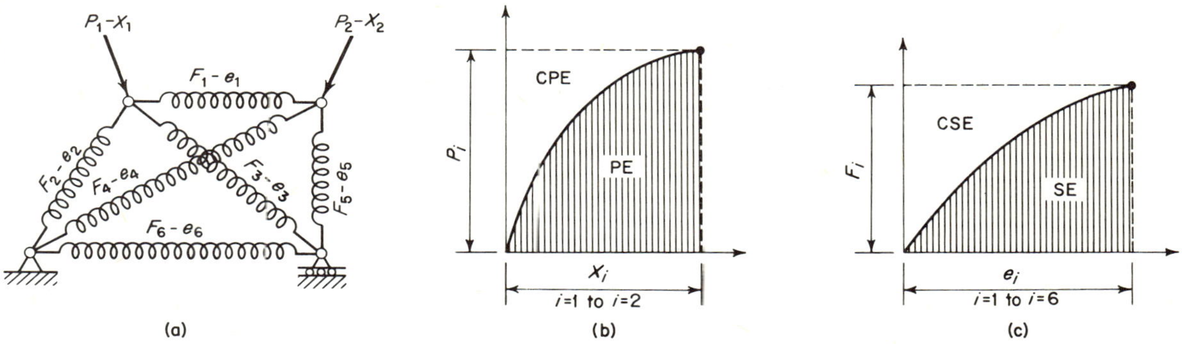

FIG. 8.11. A Parent Equilibrium and Compatible State.

(8.8.1) Σ(2 areas under 2 P–X curves) = Σ(6 areas under 6 F–e curves)

The external work done on a body has also been called *potential energy*; and the internal work stored within the body, *strain energy*. They are equal; each of them has its own physical meaning; and they are shown by the vertically hatched areas in Fig. 8.11b and c.

The areas to the left of the P–X or F–e curves in Fig. 8.11b and c, each bounded by the vertical axis and the horizontal line through the big dot, have no physical meaning, but are the complementary parts of the rectangles and have the same dimensional units as work and energy. The term *complementary-potential energy* is used to represent the sum of the areas to the left of the P–X curves; the term *complementary-strain energy*, the sum of the areas to the left of the F–e curves. It should be noted that the total complementary-potential energy need not be equal to the total complementary-strain energy. The usefulness of the concept of complementary energy will be explained in Art. 8.10.

8.9 Castigliano's First Theorem

Castigliano* in 1879 published two theorems dealing with linear structures; later Engesser** in 1889 generalized these theorems and made them applicable in both linear and nonlinear cases.

Castigliano's First Theorem may be stated as: "The partial derivative of the potential energy, or of the strain energy, with respect to a displacement at any point of a structure, is equal to the force in the direction of the displacement"; or, in symbolic form,

(8.9.1)
$$\frac{\partial(PE)}{\partial X_i} = \frac{\partial(SE)}{\partial X_i} = P_i$$

**History of Strength of Materials*, by S. Timoshenko, McGraw-Hill Book Co., 1953, p. 290.

**Ibid., p. 292.

The proof for this theorem may be made as follows: Consider the cantilever beam subjected to two P-forces as shown in Fig. 8.12a, whose parent state is shown in Fig. 8.12c and d. Define the instantaneous structure stiffness matrix $[K']$ (condensed from the more general 4 by 4 matrix involving both slopes and deflections) as

$[K'] =$

ΔP \ ΔX	1	2
1	K'_{11}	K'_{12}
2	K'_{21}	K'_{22}

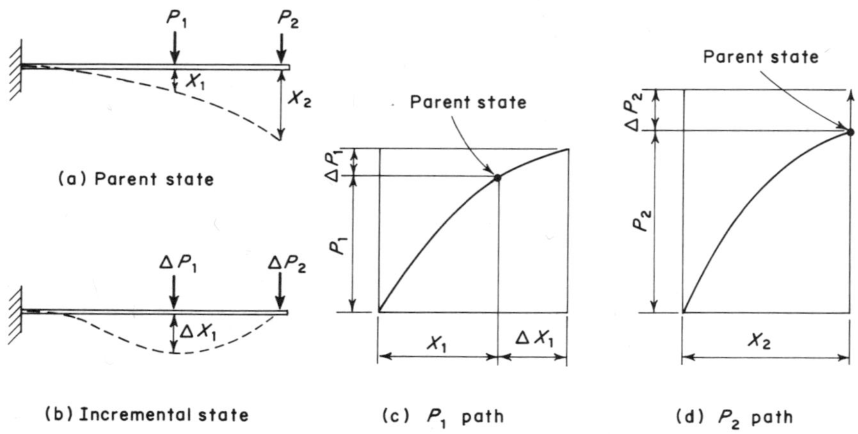

FIG. 8.12. Castigliano's First Theorem.

Let the incremental state of Fig. 8.12b be added to the parent state of Fig. 8.12a. The incremental potential energy $\Delta(PE)$ may be observed from Fig. 8.12c and d to be

$$\Delta(PE) = P_1(\Delta X_1)$$

from which

$$\frac{\partial(PE)}{\partial X_1} = P_1 \tag{8.9.2}$$

The incremental complementary-potential energy $\Delta(CPE)$ may also be observed from Fig. 8.12c and d to be

$$\Delta(CPE) = X_1(\Delta P_1) + X_2(\Delta P_2)$$

Since only $\Delta X_1 \neq 0$,

$$\Delta(CPE) = X_1(K'_{11}\Delta X_1) + X_2(K'_{21}\Delta X_1)$$

from which

$$\frac{\partial(CPE)}{\partial X_1} = K'_{11}X_1 + K'_{21}X_2 \tag{8.9.3}$$

Equation (8.9.2) is in fact identical to Eq. (8.9.1) if one can imagine the subscript "1" to be the subscript i, and at the same time generalize that there may be many other P-forces but the incremental displacement is nonzero only at the ith coordinate. Equation (8.9.3) is not in itself useful, but it will become Eq. (8.9.2) in a linear structure because then the complementary-potential energy (CPE) would be equal to the potential energy (PE) and the elements K'_{11} and K'_{21} should be invariant so as to make the right-hand side of Eq. (8.9.3) equal to P_1 in the parent state.

Castigliano's First Theorem, as represented by Eq. (8.9.1), is useful in obtaining the stiffness matrix of a structural element, first by expressing the strain energy within the element in terms of the joint displacements at the degree of freedom; then, a partial differentiation with respect to X_i will give the force P_i in terms of all the joint displacements.

The strain energy in a linear beam element can be expressed in terms of the bending moment by Eq. (8.2.3) as

$$SE = \tfrac{1}{2} \int \frac{M^2 dx}{EI}$$

or, in terms of the second derivative of the elastic curve as

$$SE = \tfrac{1}{2} \int EI\left(\frac{d^2y}{dx^2}\right)^2 dx \tag{8.9.4}$$

Example 8.4. For the cantilever beam of Fig. 8.13a, obtain expressions for M and P as functions of θ and Δ by assuming that the equation of the elastic curve is cubic (which is in fact true) and then applying Castigliano's First Theorem.

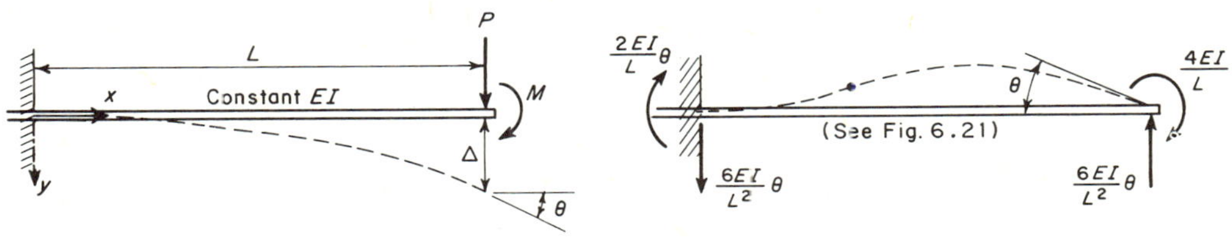

(a) (b) First mode: $\theta \neq 0, \Delta = 0$

	θ	Δ
M	$+\frac{4EI}{L}$	$-\frac{6EI}{L^2}$
P	$-\frac{6EI}{L^2}$	$+\frac{12EI}{L^3}$

(d) Stiffness matrix

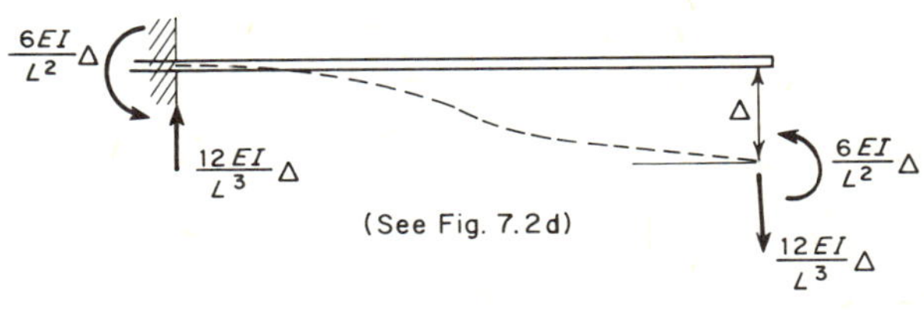

(c) Second mode: $\theta = 0, \Delta \neq 0$

FIG. 8.13. The Stiffness Matrix of a Cantilever Beam.

Solution. Actually the solution to this problem, as shown by Fig. 8.13d, can be readily assembled from Fig. 6.21 and Fig. 7.23d, reproduced here as Fig. 8.13b and c. Note that the first and second modes shown in Fig. 8.13b and c correspond to the first and second columns, respectively, in the stiffness matrix of Fig. 8.13d. Also note the strict adherence to the sign convention adopted in Fig. 8.13a where P–Δ is positive if downward, and M–θ is positive if clockwise.

If Castigliano's First Theorem is to be used, one must express the strain energy in terms of θ and Δ; or, for this matter, the four arbitrary constants a_1 to a_4 in the cubic equation of the elastic curve

$$y = a_1 + a_2x + a_3x^2 + a_4x^3$$

must be expressed in terms of θ and Δ. Using the four boundary conditions in Fig. 8.13a which are: 1) when $x = 0$, $y = 0$, 2) when $x = 0$, $dy/dx = 0$, 3) when $x = L$, $y = \Delta$, and 4) when $x = L$, $dy/dx = \theta$, the four constants a_1 to a_4 may be expressed in terms of θ and Δ as:

$$a_1 = 0$$

$$a_2 = 0$$

$$a_3 = -\frac{\theta}{L} + \frac{3\Delta}{L^2}$$

$$a_4 = +\frac{\theta}{L^2} - \frac{2\Delta}{L^3}$$

The strain energy in the cantilever beam may be obtained by using Eq. (8.9.4); thus,

$$SE = \frac{EI}{2}\int_0^L \left(\frac{d^2y}{dx^2}\right)^2 dx = \frac{EI}{2}\int_0^L (2a_3 + 6a_4x)^2\, dx$$

$$= \frac{EI}{2}\left[\frac{4\theta^2}{L} - \frac{12\theta\Delta}{L^2} + \frac{12\Delta^2}{L^3}\right]$$

Using Castigliano's First Theorem,

$$M = \frac{\partial(SE)}{\partial\theta} = +\frac{4EI\theta}{L} - \frac{6EI\Delta}{L^2}$$

$$P = \frac{\partial(SE)}{\partial\Delta} = -\frac{6EI\Delta}{L^2} + \frac{12EI\Delta}{L^3}$$

These results are identical to those of Fig. 8.13d, which have been obtained previously as in Fig. 8.13b and c by the physical-geometric (i.e., moment-area) method.

EXERCISE 8.3

Express the clockwise end moments M_i and M_j which should be applied to a prismatic beam with two hinged ends in terms of the clockwise end rotations ϕ_i and ϕ_j, by assuming that the equation of the elastic curve is cubic (which is in fact true) and then applying Castigliano's First Theorem.

8.10 Castigliano's Second Theorem

Castigliano's Second Theorem may be stated as: "The partial derivative of the complementary-potential energy, or of the complementary-strain energy, with respect to a force at any point of a structure, is equal to the displacement in the direction of the force"; or, in symbolic form,

$$\frac{\partial(CPE)}{\partial P_i} = \frac{\partial(CSE)}{\partial P_i} = X_i \tag{8.10.1}$$

The proof of this theorem may be made as follows: Consider the cantilever beam subjected to two P-forces as shown in Fig. 8.14a, whose parent state is shown in Fig. 8.14c and d. Define the instantaneous structure flexibility matrix $[\delta']$ as

$[\delta'] =$

ΔX \ ΔP	1	2
1	δ'_{11}	δ'_{12}
2	δ'_{21}	δ'_{22}

Let the incremental state of Fig. 8.14b be added to the parent state of Fig. 8.14a. The incremental potential energy $\Delta(PE)$ may be observed from Fig. 8.14c and d to be

$$\Delta(PE) = P_1(\Delta X_1) + P_2(\Delta X_2)$$

Since only $\Delta P_1 \neq 0$,

$$\Delta(PE) = P_1(\delta'_{11}\Delta P_1) + P_2(\delta'_{21}\Delta P_1)$$

from which

$$\frac{\partial(PE)}{\partial P_1} = \delta'_{11}P_1 + \delta'_{21}P_2 \tag{8.10.2}$$

The incremental complementary-potential energy may also be observed from

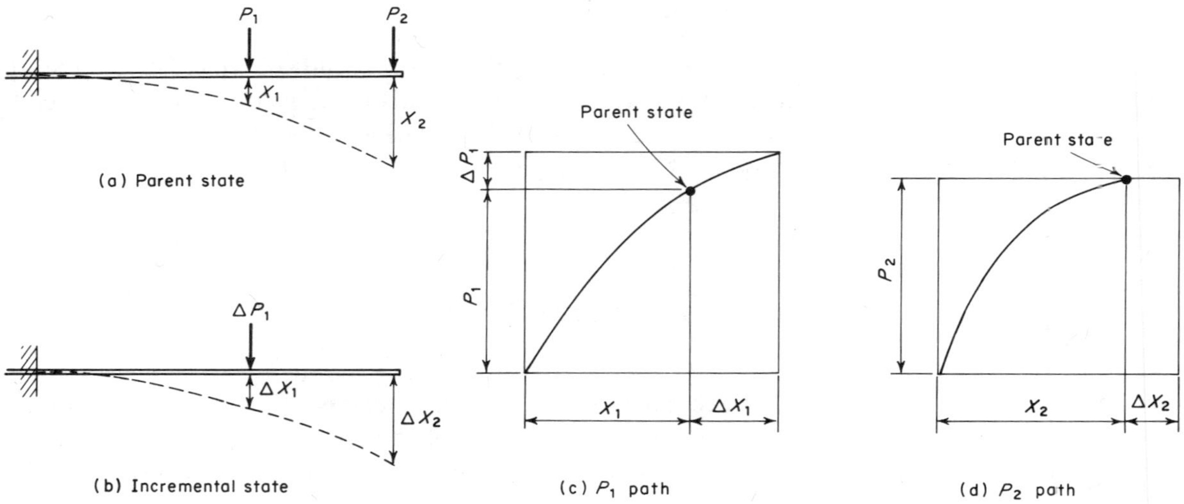

FIG. 8.14. Castigliano's Second Theorem.

Fig. 8.14c and d to be

$$\Delta(CPE) = X_1(\Delta P_1)$$

from which

$$\frac{\partial(CPE)}{\partial P_1} = X_1 \tag{8.10.3}$$

Equation (8.10.3) is in fact identical to Eq. (8.10.1), if one can imagine the subscript "1" to be the subscript i, and at the same time generalize that there may be many other P-forces but the incremental force is nonzero only at the ith coordinate. Equation (8.10.2) is not useful when taken by itself, but it will become Eq. (8.10.3) in a linear structure because the potential energy (PE) then would be equal to the complementary potential energy (CPE) and the elements δ'_{11} and δ'_{21} should be invariant so as to make the right-hand side of Eq. (8.10.2) equal to X_1 in the parent state.

Castigliano's Second Theorem, as represented by Eq. (8.10.1), is useful in obtaining the flexibility matrix of a structural element first by expressing the complementary-strain energy within the element in terms of the forces at the degree of freedom; then, a partial differentiation with respect to P_i will give the displacement X_i in terms of all the forces.

The complementary-strain energy in a *linear* beam element has to be equal to the strain energy; thus, Eq. (8.2.3) may be restated as

$$SE = CSE = \tfrac{1}{2}\int \frac{M^2dx}{EI} \tag{8.10.4}$$

Example 8.5. Express the clockwise rotations ϕ_i and ϕ_j at the ends of a prismatic beam with two hinged ends in terms of the clockwise end moments M_i and M_j by applying Castigliano's Second Theorem.

Solution. Actually the solution to this problem, as shown by Fig. 8.15d, can be readily assembled from the results of the application of Conjugate-Beam Theorem I to the simple beams shown in Fig. 8.15b and c. Note that the first and second modes shown in Fig. 8.15b and c correspond to the first and second columns in the flexibility matrix of Fig. 8.15d. Also note the strict adherence to the sign convention adopted in Fig. 8.15a wherein both pairs of M_i-ϕ_i and M_j-ϕ_j are positive if clockwise.

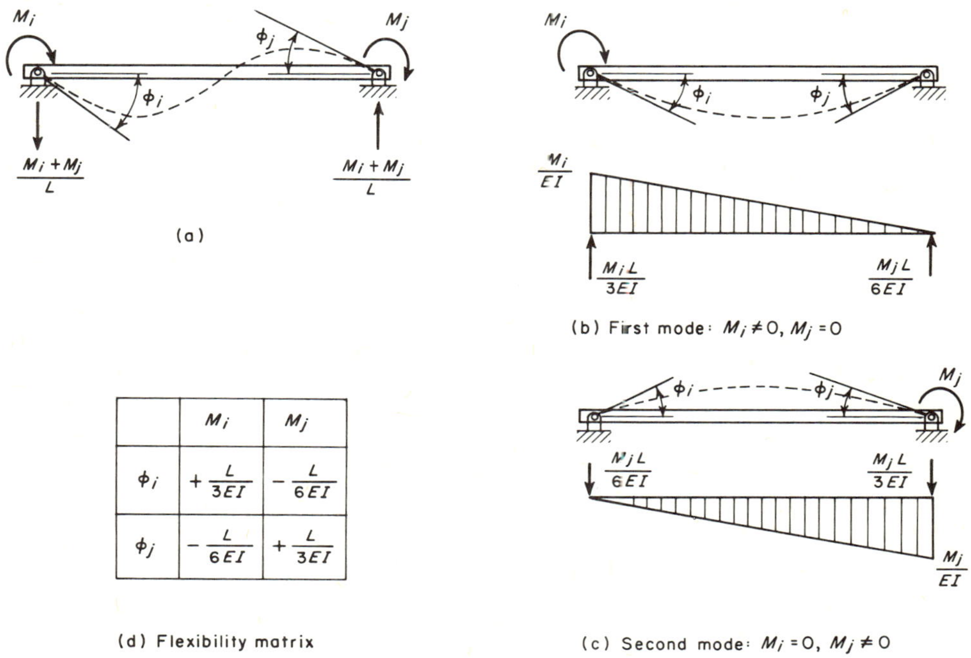

	M_i	M_j
ϕ_i	$+\frac{L}{3EI}$	$-\frac{L}{6EI}$
ϕ_j	$-\frac{L}{6EI}$	$+\frac{L}{3EI}$

(d) Flexibility matrix

FIG. 8.15. The Flexibility Matrix of a Simple Beam.

Although Castigliano's Second Theorem calls for the complementary energy instead of the "regular" energy, they are the same for a linear structure; thus, referring to Fig. 8.15a,

$$CSE = SE = \tfrac{1}{2}\int \frac{M^2 dx}{EI} = \frac{1}{2EI}\int_0^L \left(M_i - \frac{M_i + M_j}{L}x\right)^2 dx$$
$$= \frac{M_i^2 L}{6EI} - \frac{M_i M_j L}{6EI} + \frac{M_j^2 L}{6EI}$$

Using Castigliano's Second Theorem,

$$\phi_i = \frac{\partial(CSE)}{\partial M_i} = +\frac{M_i L}{3EI} - \frac{M_j L}{6EI}$$
$$\phi_j = \frac{\partial(CSE)}{\partial M_j} = -\frac{M_i L}{6EI} - \frac{M_j L}{3EI}$$

These results are identical to those of Fig. 8.15d, which have been previously obtained by the physical-geometric (i.e., conjugate-beam) method.

EXERCISE 8.4

By means of Castigliano's Second Theorem, express the clockwise rotation θ and the downward deflection Δ at the free end of a prismatic cantilever beam in terms of a clockwise moment M and a downward force P, both applied simultaneously at the free end.

8.11 Derivation of the Unit-Load Method from Castigliano's Second Theorem

Chronologically the unit-load method of finding truss deflections was first proposed by Maxwell* in 1864 but remained unnoticed until later rediscovered independently by Mohr** in 1874. The generalization of the unit-load method from truss to beams and rigid frames was made by Castigliano*** in 1879 through his second theorem.

As mentioned earlier in Art. 8.9, Castigliano in 1879 used the partial derivative of the "regular" potential or strain energy in both of his theorems; Engesser in 1889 pointed out that, for nonlinear structures, the use of the "regular" energy was still correct in the first theorem, but the complementary energy should be employed in the second theorem.

The unit-load method for finding deflections in *linear* trusses may be derived from Castigliano's Second Theorem as follows:

$$X_i = \frac{\partial(CSE)}{\partial P_i} = \frac{\partial(SE)}{\partial P_i} = \frac{\partial}{\partial P_i}\left(\Sigma\frac{F^2L}{2AE}\right) = \Sigma\left(\frac{FL}{AE}\right)\left(\frac{\partial F}{\partial P_i}\right)$$

$$= \Sigma\begin{pmatrix}e \text{ in real compat-}\\ \text{ible state}\end{pmatrix}\begin{pmatrix}F \text{ due to unit load}\\ \text{in equilibrium state}\end{pmatrix}$$

which is Eq. (3.6.2). Likewise, the unit-load method of finding deflections and slopes in *linear* beams and rigid frames may be derived:

$$X_i = \frac{\partial(CSE)}{\partial P_i} = \frac{\partial(SE)}{\partial P_i} = \frac{\partial}{\partial P_i}\left(\int\frac{M^2dx}{2EI}\right) = \int\frac{M(\partial M/\partial P_i)\,dx}{EI}$$

$$= \int\frac{Mmdx}{EI}$$

which is Eqs. (5.7.6) or (5.7.7).

***History of Strength of Materials*, by S. P. Timoshenko, McGraw-Hill Book Company, 1953, p. 202.

**Ibid., p. 207.

***Ibid., p. 292.

8.12 The Theorem of Least Work

The Theorem of Least Work may be stated as: "The redundant forces in a statically indeterminate structure are such as to make the total complementary-strain energy a minimum." When reference is made only to linear structure, the "regular" strain energy can be used in place of the complementary-strain energy.

The redundant forces in a statically indeterminate structure have been defined in Art. 4.3 for trusses, in Arts. 6.4 and 6.5 for beams, and in Art. 7.4 for rigid frames. In the force method of analysis, these redundant forces must be solved first from the compatibility equations of deformation, in one-to-one correspondence with the redundant forces. It will be shown that the Theorem of Least Work will yield a set of equations identical to the compatibility equations in the force method.

The validity of the Theorem of Least Work may be established by considering the typical structures in Fig. 8.16. The statically indeterminate beam of Fig. 8.16a may be considered to be a degenerate statically determinate beam of Fig. 8.16b subjected to the applied loads and the unknown redundant forces p_1

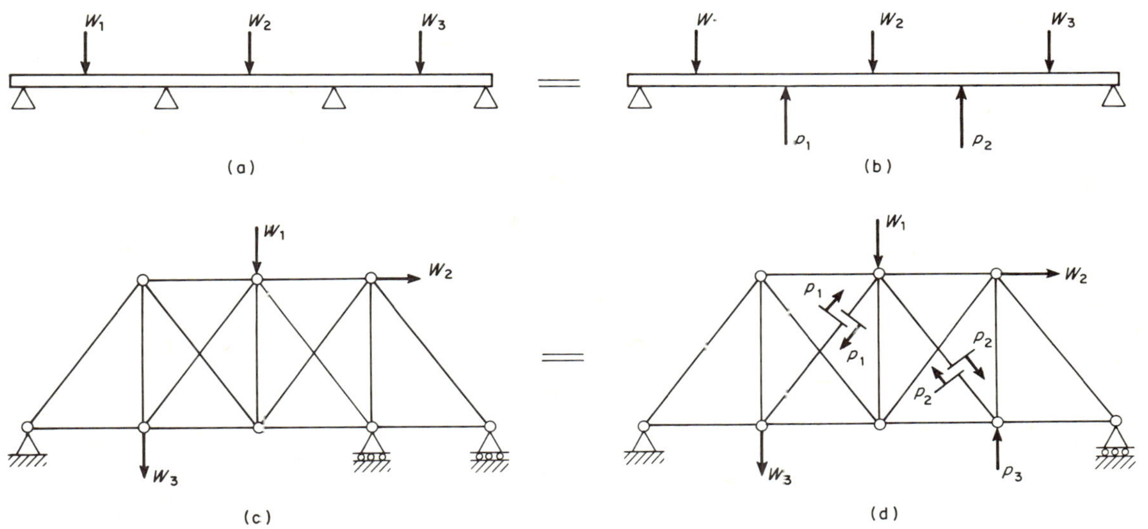

FIG. 8.16. The Theorem of Least Work.

and p_2. The statically indeterminate truss of Fig. 8.16c may be considered to be a degenerate statically determinate truss of Fig. 8.16d subjected to the applied loads and the unknown redundant forces p_1, p_2, and p_3 wherein the pairs of colinear forces p_1-p_2 and p_2-p_2 are applied as pulling forces at the ends of the cut members. Since the deflection x_i in the direction of a redundant reacting force p_i at a rigid support, or the overlap distance x_i in the direction of the pair of colinear pulling forces p_i-p_i acting at the cut ends should be zero, applying

Castigliano's Second Theorem gives

$$x_i = \frac{\partial(CSE)}{\partial p_i} = 0 \text{ (for } i = 1 \text{ to } i = \text{degree of indeterminacy)} \tag{8.12.1}$$

which is the Theorem of Least Work.

Because x_i is the deflection at the ith coordinate in the parent condition (which is the real condition to be solved),

$$x_i = \Sigma\delta_{ij}p_j \text{ (for } j = 1 \text{ to } j = \text{total number of external forces acting on the equivalent degenerate structure)} \tag{8.12.2}$$

The second partial derivative of the complementary-strain energy with respect to x_i is

$$\frac{\partial^2(CSE)}{\partial p_i^2} = \frac{\partial x_i}{\partial p_i} = \delta_{ii} \tag{8.12.3}$$

which has to be positive due to the fact that in a basic mode wherein only $p_i \neq 0$, the potential energy is $\frac{1}{2}(\delta_{ii}p_i)(p_i)$.

In the force method of analysis the compatibility equations $x_i = 0$ for $i = 1$ to $i =$ degree of indeterminacy are directly used as such. Then the required deflections in the ith coordinate of the degenerate structure due to the applied loads and due to each unit value of the unknown redundant forces can be obtained by either the physical-geometric or unit-load methods. Because the use of the unit-load method within the force method was originally developed by Maxwell and Mohr (see Art. 8.11), the resulting compatibility equations of the force method are sometimes called the *Maxwell–Mohr Equations*.

In using the Theorem of Least Work for the analysis of statically indeterminate structures, the partial-derivative equations,

$$\frac{\partial(CSE)}{\partial p_i} = 0 \quad \text{(for } i = 1 \text{ to } i = \text{degree of indeterminacy)}$$

are directly used as such. For *linear* structures, the strain energy within the structure is first expressed in terms of the redundant forces; then the compatibility equations are obtained by equating with zero the partial derivatives of the strain energy with respect to each of the redundant forces. Naturally, the equations thus obtained are identical to the Maxwell-Mohr Equations. Nevertheless, the Least-Work Theorem is the most elegant, and perhaps the most important concept in the entire subject of "Theory of Statically Indeterminate Structures."

Example 8.6. Establish, by the Theorem of Least Work, the compatibility equation obtained for the statically indeterminate truss of Example 4.1, shown again in Fig. 8.17a.

Solution. Using the force in member L_1U_2 as redundant, the compatibility

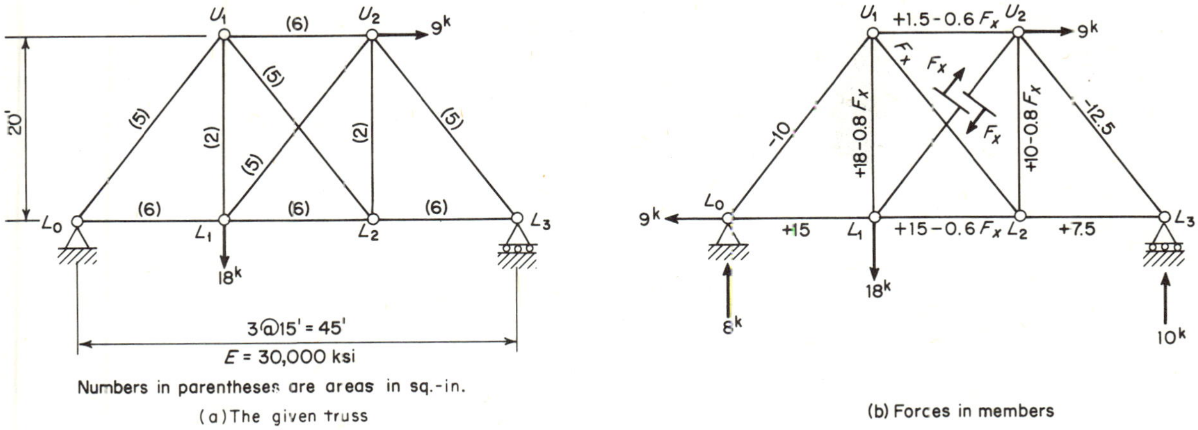

FIG. 8.17. Least-Work Theorem Solution (Example 8.6).

equation has been obtained by the force method in Example 4.1 to be

$$-124.5 + 9.84F_x = 0$$

By the Theorem of Least Work, the compatibility equation is

$$\frac{\partial(CSE)}{\partial F_x} = \frac{\partial(SE)}{\partial F_x} = \frac{\partial}{\partial F_x}\left[\Sigma\frac{F^2L}{2AE}\right] = \Sigma\left[\frac{FL}{AE}\left(\frac{\partial F}{\partial F_x}\right)\right] = 0$$

The expressions for F in Table 8.1 are taken from Fig. 8.17b; these forces are in fact the superposition of the forces due to the applied loads and due to the

TABLE 8.1. Least-Work Theorem Solution—Example 8.6

No.	Member	F	$\frac{\partial F}{\partial F_x}$	$\frac{L}{EA}(10^{-3}$ in.)	$\frac{FL}{EA}\left(\frac{\partial F}{\partial F_x}\right)$
1	U_1U_2	$+1.5 - 0.6F_x$	-0.6	1.0	$-0.9 + 0.36F_x$
2	L_0L_1	$+15.0$		1.0	
3	L_1L_2	$+15.0 - 0.6F_x$	-0.6	1.0	$-9.0 + 0.36F_x$
4	L_2L_3	$+7.5$		1.0	
5	U_1L_1	$+18.0 - 0.8F_x$	-0.8	4.0	$-57.6 + 2.56F_x$
6	U_2L_2	$+10.0 - 0.8F_x$	-0.8	4.0	$-32.0 + 2.56F_x$
7	L_0U_1	-10.0		2.0	
8	U_1L_2	$-12.5 + 1.0F_x$	$+1.0$	2.0	$-25.0 + 2.00F_x$
9	L_1U_2	$+1.0F_x$	$+1.0$	2.0	$+2.00F_x$
10	U_2L_3	-12.5		2.0	
Total					$-124.5 + 9.84F_x$

pair of redundant forces F_x and F_x. Note that in the present solution, the redundant bar L_1U_2 is merely cut, but not removed; hence, it should be included in the tabulation just like the rest of the bars. When the redundant bar is removed as in Example 4.1, the compatibility equation is established by the physical requirement that the relative displacement apart between joints L_1 and U_2 must be equal to the elongation of the bar.

Example 8.7. Establish the compatibility equation obtained for the statically indeterminate rigid frame of Example 7.3, shown again in Fig. 8.18a, by the Theorem of Least Work.

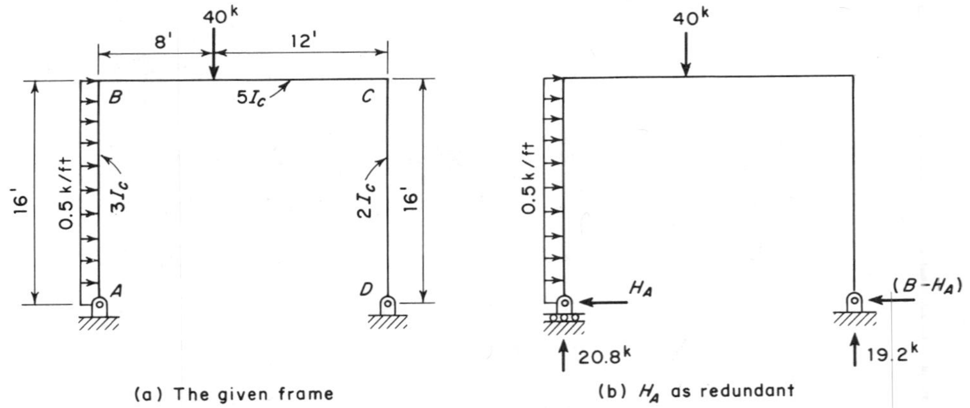

FIG. 8.18. Least-Work Theorem Solution (Example 8.7).

Solution. Using the horizontal reaction H_A as redundant, the compatibility equation has been obtained by the force method in Example 7.3 to be

$$\frac{20480}{3} - \frac{19456}{9}H_A = 0$$

By the Theorem of Least Work, the compatibility equation is

$$\frac{\partial(CSE)}{\partial H_A} = \frac{\partial(SE)}{\partial H_A} = \frac{\partial}{\partial H_A}\left[\Sigma \int \frac{M^2\,dx}{2EI}\right] = 0$$

Referring to Fig. 8.18b,

$$SE = \Sigma \int \frac{M^2\,dx}{2EI} = \frac{1}{2E(3I_c)}\int_0^{16}\left[H_A x - \frac{0.5x^2}{2}\right]^2 dx$$
$$+ \frac{1}{2E(5I_c)}\int_0^{8}[20.8x - 64 + 16H_A]^2\,dx$$
$$+ \frac{1}{2E(5I_c)}\int_0^{12}[19.2x - (8 - H_A)(16)]^2\,dx$$
$$+ \frac{1}{2E(2I_c)}\int_0^{16}[(8 - H_A)x]^2\,dx$$

$$
\begin{aligned}
\frac{\partial(SE)}{\partial H_A} &= \frac{1}{E(3I_c)}\int_0^{16}\left[H_A x - \frac{0.5x^2}{2}\right](x)\,dx \\
&\quad + \frac{1}{E(5I_c)}\int_0^{8}[20.8x - 64 + 16H_A](16)\,dx \\
&\quad + \frac{1}{E(5I_c)}\int_0^{12}[19.2x - 12.8 + 16H_A](16)\,dx \\
&\quad + \frac{1}{E(2I_c)}\int_0^{16}[8x - H_A x](-x)\,dx \\
&= \left[\frac{4096}{9}H_A - \frac{4096}{3}\right] + [+491.52 - 409.6H_A] + [-491.52 + 614.4H_A] \\
&\quad + \left[-\frac{16384}{3} + \frac{2048}{3}H_A\right] \\
&= \frac{19456}{9}H_A - \frac{20480}{3} = 0
\end{aligned}
$$

EXERCISE 8.5

Establish the compatibility equation of the truss in Exercise 4.1 using the force in bar L_1U_2 as the redundant by the Theorem of Least Work.

EXERCISE 8.6

Establish the compatibility equation of the rigid frame in Exercise 7.3 using the horizontal reaction at D as the redundant by the Theorem of Least Work.

8.13 Concluding Remarks

The important energy theorems in structural analysis have been assembled and described in this chapter. Only after the material in the first seven chapters has been thoroughly assimilated, is it possible that these theorems may be explained by drawing upon some past experience.

Two distinct groups of work and energy theorems have been presented: those relating to real work, virtual work, transposition relationship and reciprocal theorems in Arts. 8.2 to 8.6; and those relating to the two Castigliano theorems in Arts. 8.8 to 8.12. Except the Principle of Real Work, the theorems in the first group apply only to linear structures or the linearized increments of nonlinear structures, but those in the second group are applicable to linear as well as nonlinear structures.

Castigliano's First Theorem is useful in establishing the stiffness matrix; Castigliano's Second Theorem, the flexibility matrix. The former is related to the displacement method of analysis; the latter to the force method of analysis. Up to this day, the two Castigliano theorems, first stated in 1879, constitute the foundation for the Theory of Structures.

Influence Lines and Moving Live Loads

9

9.1 Fixed and Moving Live Loads

Fixed loads on structures are those of constant magnitudes at fixed locations. Examples are the dead weight of the structure itself, and weights of permanent installations as walls, machinery, or equipment. Often structures are built to carry loads such as people, furniture, fluid pressure, or traffic; these loads are moving live loads because they may or may not come within a range of positions. The determination of the critical position of any one live load or combination of several live loads that causes a maximum external reaction or internal force, which may be the axial force in a bar of a truss or the bending moment at a section in a rigid frame, is a subject of vital importance in structural analysis.

It will be shown that the construction of influence lines is the key to the solution for the critical positions of moving live loads.

9.2 Influence Lines for Reactions on Statically Determinate Beams

The influence line for a function is the graph showing the value of that function owing to a single concentrated load moving across a structure, wherein the value of the function is plotted directly at the location of the single load. Influence lines for statically determinate structures are composed of straight segments because the functions (external reactions or internal forces) are always linear with respect to the position of the single concentrated load.

The influence lines for the two reactions R_A and R_B on the overhanging beam of Fig. 9.1a are shown in Fig. 9.1b and c; and those for the two reactions

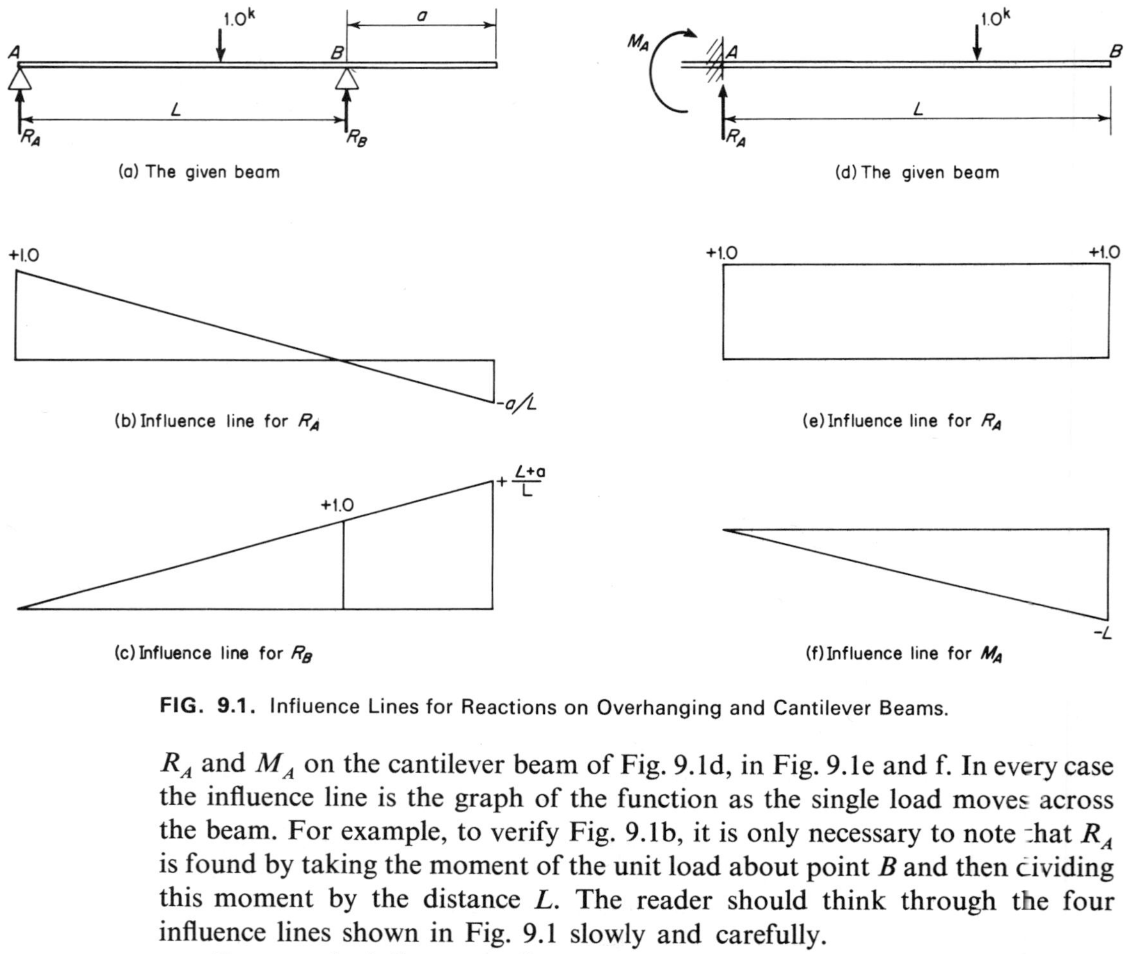

FIG. 9.1. Influence Lines for Reactions on Overhanging and Cantilever Beams.

R_A and M_A on the cantilever beam of Fig. 9.1d, in Fig. 9.1e and f. In every case the influence line is the graph of the function as the single load moves across the beam. For example, to verify Fig. 9.1b, it is only necessary to note that R_A is found by taking the moment of the unit load about point B and then dividing this moment by the distance L. The reader should think through the four influence lines shown in Fig. 9.1 slowly and carefully.

Because the influence ordinate at any location is the value of the function due to a unit load at that location, the value of the function due to the *influence* of several loads such as W_1-W_2-W_3 in Fig. 9.2a and b should be

$$R_A = W_1y_1 + W_2y_2 + W_3y_3 \tag{9.2.1}$$

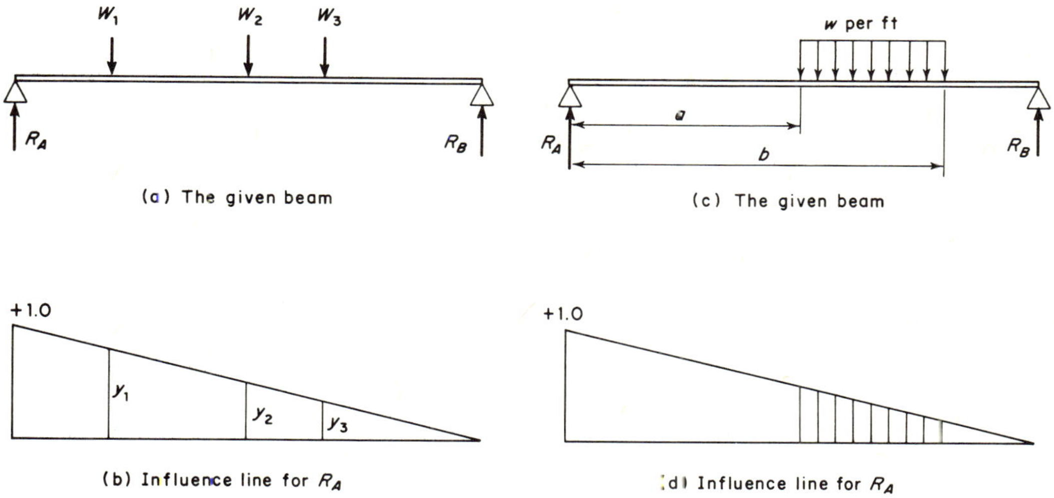

FIG. 9.2. Computation for Functions by Influence Lines.

Here one can consider either W_1-W_2-W_3 to be the dimensionless ratios with y_1-y_2-y_3 in kips, or W_1-W_2-W_3 in kips with y_1-y_2-y_3 as dimensionless ratios. Further, the value of the function due to a given length of uniform load such as w in Fig. 9.2c and d is

$$R_A = \int_a^b (w\,dx)(y) = w \int_a^b y\,dx \tag{9.2.2}$$

Here one can consider either w as a dimensionless ratio per foot and the area of the influence diagram under the uniform load in ft-kips, or w in kips per ft and the influence ordinates as dimensionless ratios.

The influence line is first useful in determining, by visual inspection in most cases, the critical position of a series of moving concentrated loads of known magnitudes at known spacings, or of a moving uniform load of definite or indefinite length. Once the critical position has been obtained, the critical value of the function itself may be computed either by using Eqs. (9.2.1) and (9.2.2) or by the ordinary methods of structural analysis for fixed loads. The suggestion is to do this computation by both ways in order to see that the same results are obtained.

Example 9.1. Construct the influence line for R_B on the overhanging beam of Fig. 9.3a; and determine the maximum positive and maximum negative values of R_B due to the passage: (a) of the three concentrated loads of Fig. 9.3c in either direction; and (b) of the uniform load of indefinite length of Fig. 9.3d.

Solution. The influence line for R_B is shown in Fig. 9.3b.

(a) *Concentrated loads.* The critical position for the maximum positive value of R_B is visually determined and shown in Fig. 9.3e; its value is computed by

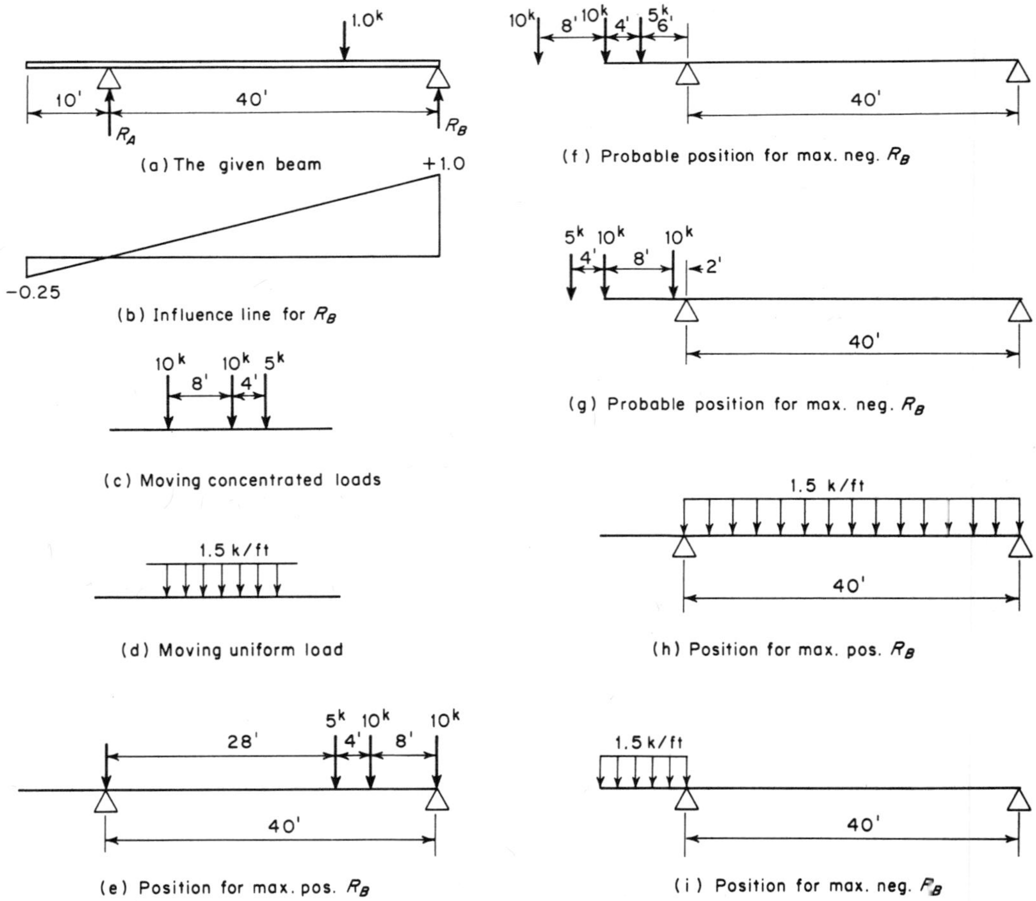

FIG. 9.3. Maximum Reaction on an Overhanging Beam.

Eq. (9.2.1) to be

$$\text{Max. pos. } R_B = \Sigma Wy = 10(1.0) + 10(32/40) + 5(28/40)$$
$$= 10 + 8 + 3.5 = +21.5 \text{ kips}$$

or by considering the free body of Fig. 9.3e,

$$\text{Max. pos. } R_B = \frac{\text{moments about } A}{40} = \frac{5(28) + 10(32) + 10(40)}{40}$$
$$= \frac{860}{40} = +21.5 \text{ kips (check)}$$

Since the critical position for the maximum negative value of R_B could be either Fig. 9.3f or g, it is necessary to compute its value in both cases. These values are -3.25 and -3.00 kips respectively. Thus Fig. 9.3f gives the

$$\text{Max. neg. } R_B = -3.25 \text{ kips}$$

(b) *Uniform load.* From Fig. 9.3h,

$$\text{Max. pos. } R_A = \frac{1.5(40)}{2} = +30 \text{ kips}$$

and, from Fig. 9.3i,

$$\text{Max. neg. } R_A = -\frac{1.5(10)^2/2}{40} = -\frac{75}{40} = -1.875 \text{ kips}$$

The above values may be verified by using Eq. (9.2.2).

EXERCISE 9.1

Referring to Fig. 9.4 construct the influence lines for R_B and M_B acting on the cantilever beam, and determine their maximum positive and maximum negative values due to the passage: (a) of the three concentrated loads in either direction; and (b) of the uniform load of indefinite length.

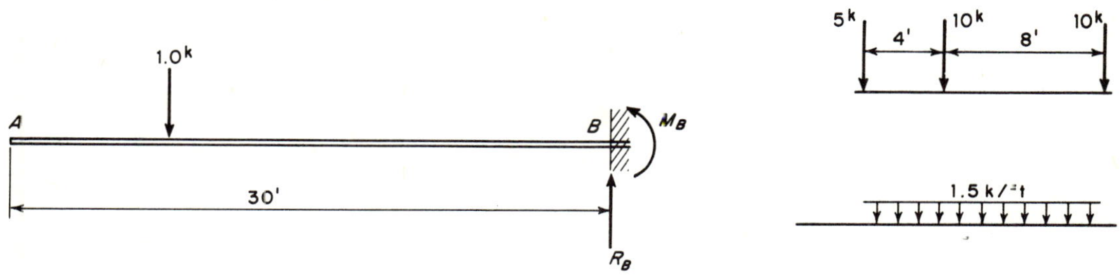

FIG. 9.4. Exercise 9.1.

9.3 Influence Lines for Shear and Bending Moment in Statically Determinate Beams

The influence lines for shear and bending moment at a chosen section in a statically determinate beam may be constructed by computing the influence ordinates at the critical positions of the unit load, and then connecting the critical points by straight lines. For instance, the critical ordinates on the influence line in Fig. 9.5b for shear at section C of the overhanging beam of Fig. 9.5a are

1) for a unit load at A,

$$V_C = -1 + R_B = -1 + \frac{a - L}{L} = +\frac{a}{L}$$

2) for a unit load at B,

$$V_C = -1 + R_B = -1 + 1 = 0$$

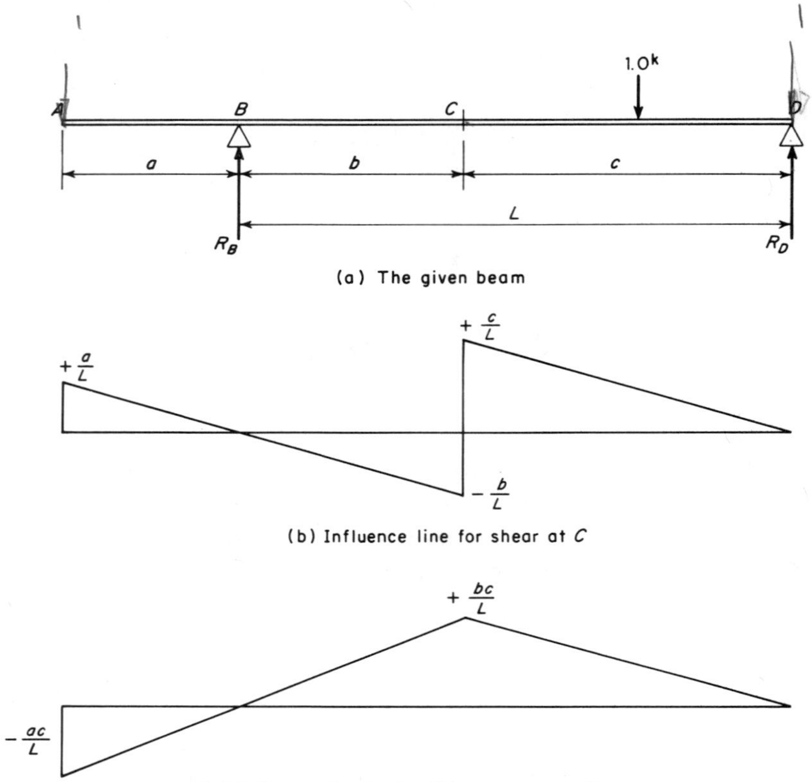

(c) Influence line for bending moment at C

FIG. 9.5. Influence Lines for Shear and Bending Moment.

3) for a unit load at a section just to left of C,

$$V_C = R_B - 1 = \frac{c}{L} - 1 = -\frac{b}{L}$$

4) for a unit load at a section just to right of C,

$$V_C = R_B = +\frac{c}{L}$$

5) for a unit load at D,

$$V_C = R_B = 0$$

The critical ordinates on the influence line (Fig. 9.5c) for bending moment at section C are

1) for a unit load at A,

$$M_C = R_D(c) = \left(-\frac{a}{L}\right)(c) = -\frac{ac}{L}$$

2) for a unit load at B,

$$M_C = 0$$

3) for a unit load at C,

$$M_C = R_B(b) = \left(\frac{c}{L}\right)(b) = +\frac{bc}{L}$$

4) for a unit load at D,

$$M_C = 0$$

Example 9.2. Construct the influence lines for shear and bending moment at section B of the overhanging beam $ABCD$ shown in Fig. 9.6a; compute the maximum and minimum shears and bending moments at B due to a dead load of 0.6 kip per ft and a live load of 1.5 kips per ft, assuming that the live load may be broken into segments of any length.

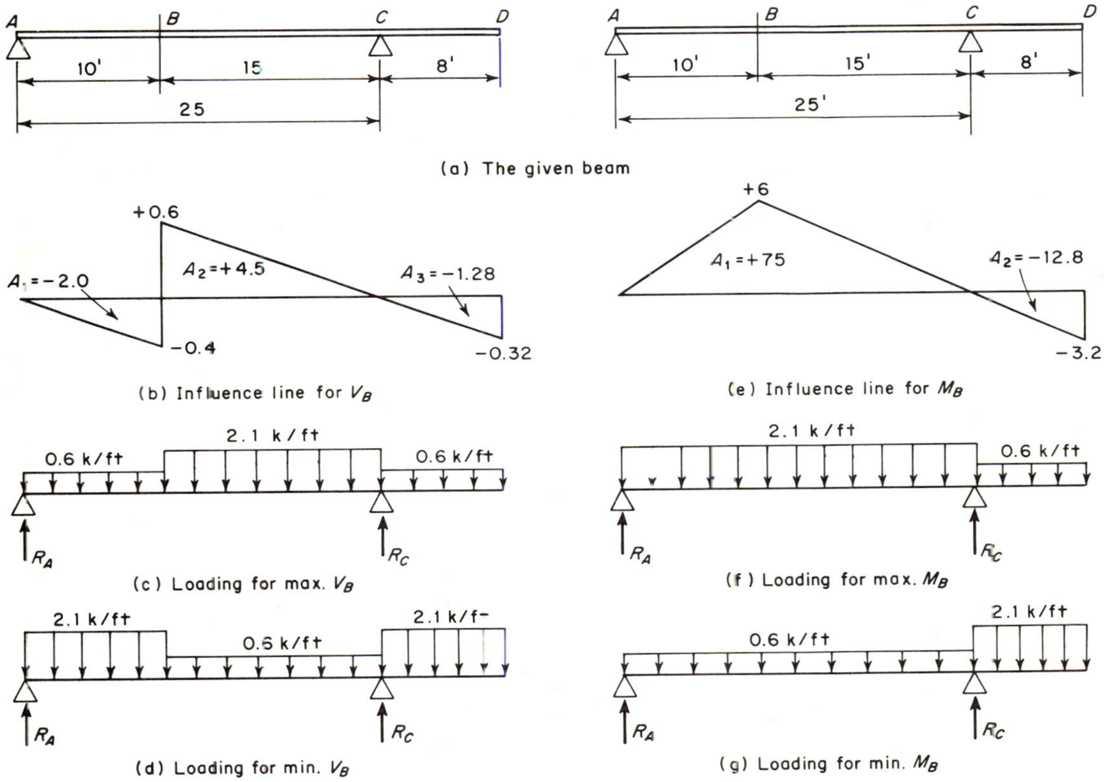

FIG. 9.6. Maximum and Minimum Shears and Bending Moments.

Solution

(a) *Maximum and minimum shears at B.* The influence line for V_B is constructed as shown in Fig. 9.6b; its component areas A_1, A_2, and A_3 are

$$A_1 = \tfrac{1}{2}(-0.4)(10) = -2.0$$
$$A_2 = \tfrac{1}{2}(+0.6)(15) = +4.5$$
$$A_3 = \tfrac{1}{2}(-0.32)(8) = -1.28$$

The loading pattern for maximum shear at B is shown in Fig. 9.6c; its value may be obtained by using Eq. (9.2.2) as

$$\begin{aligned}\text{Maximum } V_B &= 0.6A_1 + 2.1A_2 + 0.6A_3\\ &= 0.6(-2.0) + 2.1(+4.5) + 0.6(-1.28)\\ &= +7.482 \text{ kips}\end{aligned}$$

This value should be checked by computing the shear at B of the beam of Fig. 9.6c; or

$$\begin{aligned}\text{Maximum } V_B &= R_A - 0.6(10) = \frac{0.6(33)(8.5) + 1.5(15)(7.5)}{25} - 6\\ &= 13.482 - 6 = +7.482 \text{ kips (check)}\end{aligned}$$

In similar manner, applying Eq. (9.2.2) to the loading pattern of Fig. 9.6d,

$$\begin{aligned}\text{Minimum } V_B &= 2.1A_1 + 0.6A_2 + 2.1A_3\\ &= 2.1(-2.0) + 0.6(+4.5) + 2.1(-1.28)\\ &= -4.188 \text{ kips}\end{aligned}$$

Applying the conventional method to the beam of Fig. 9.6d,

$$\begin{aligned}\text{Minimum } V_B &= R_A - 2.1(10) = \frac{0.6(33)(8.5) + 1.5(10)(20) - 1.5(8)(4)}{25} - 21.0\\ &= 16.812 - 21.0 = -4.188 \text{ kips (check)}\end{aligned}$$

(b) *Maximum and minimum bending moments at B.* The influence line for M_B is constructed as shown in Fig. 9.6e; its component areas A_1 and A_2 are

$$\begin{aligned}A_1 &= \tfrac{1}{2}(+6)(25) = +75\\ A_2 &= \tfrac{1}{2}(-3.2)(8) = -12.8\end{aligned}$$

Applying Eq. (9.2.2) to the loading pattern of Fig. 9.6f,

$$\begin{aligned}\text{Maximum } M_B &= 2.1A_1 + 0.6A_2\\ &= 2.1(+75) + 0.6(-12.8)\\ &= +149.82 \text{ ft-kips}\end{aligned}$$

Checking by using the free body shown by Fig. 9.6f,

$$\begin{aligned}\text{Maximum } M_B &= R_A(10) - \tfrac{1}{2}(2.1)(10)^2\\ &= \left[\frac{0.6(33)(8.5) + 1.5(25)(12.5)}{25}\right](10) - 105\\ &= 25.482(10) - 105 = +149.82 \text{ ft-kips (check)}\end{aligned}$$

Applying Eq. (9.2.2) to the loading pattern of Fig. 9.6g,

$$\begin{aligned}\text{Minimum } R_B &= 0.6A_1 + 2.1A_2\\ &= 0.6(+75) + 2.1(-12.8)\\ &= +18.12 \text{ ft-kips}\end{aligned}$$

Checking by using the free body shown by Fig. 9.6g,

$$\text{Minimum } R_B = R_A(10) - \tfrac{1}{2}(0.6)(10)^2$$
$$= \left[\frac{0.6(33)(8.5) - 1.5(3)(4)}{25}\right](10) - 30$$
$$= 4.812(10) - 30 = +18.12 \text{ ft-kips (check)}$$

Had there been a reversal in sign between the maximum and minimum values of M_B and had the beam been built in reinforced concrete, tensile reinforcement would be required at both top and bottom of the cross section at B.

EXERCISE 9.2

Referring to Fig. 9.7 construct the influence lines for shear and bending moment at section C of the overhanging beam $ABCD$, and compute the maximum and minimum shears and bending moments at C due to a dead load of 0.8 kip per ft and a live load of 2 kips per ft, assuming that the live load may be broken into segments of any length.

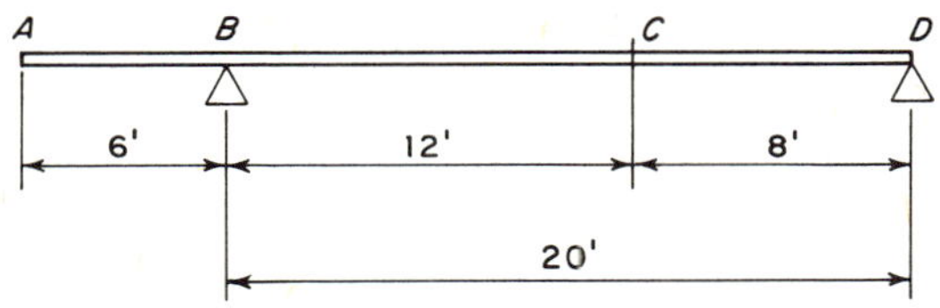

FIG. 9.7. Exercise 9.2.

9.4 Maximum Bending Moment at a Section of a Simple Beam Due to a Long Series of Moving Concentrated Loads

When a long series of concentrated loads passes over a simple beam in either direction, its critical position effecting the maximum bending moment at a chosen cross section may be obtained by trial and error with the aid of Eq. (9.2.1), but it would be far more desirable if it were possible to derive some kind of algebraic criterion.

First, it can be shown that if the influence line is a straight line segment, the value of a function due to a group of wheel loads is equal to the product of the influence ordinate at the location of the resultant force G, and the value of G itself. Referring to Fig. 9.8, and by definition of the resultant force,

$$G\bar{x} = W_1x_1 + W_2x_2 + W_3x_3 + W_4x_4 \qquad (9.4.1)$$

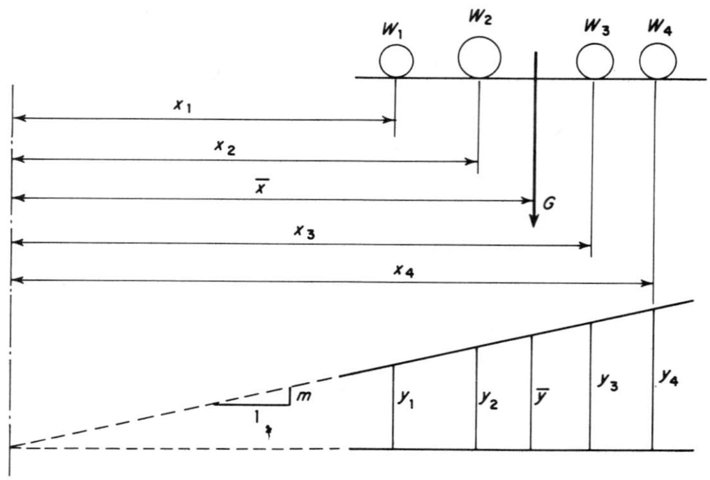

FIG. 9.8. Influence Ordinate Under a Resultant Force.

Multiplying every term in Eq. (9.4.1) by the slope m of the influence line segment,

$$Gm\bar{x} = W_1mx_1 + W_2mx_2 + W_3mx_3 + W_4mx_4$$

from which

$$G\bar{y} = W_1y_1 + W_2y_2 + W_3y_3 + W_4y_4 = \Sigma Wy \tag{9.4.2}$$

Now consider the primary wheel position of Fig. 9.9 to be advanced a distance Δx toward the left. The change in the bending moment at C is

$$\begin{aligned}\Delta M_C &= G_1(-m_1\Delta x) + G_2(+m_2\Delta x)\\ &= -\frac{G_1b}{L}\Delta x + \frac{G_2a}{L}\Delta x\\ &= \left(\frac{Ga}{L} - G_1\right)\Delta x\end{aligned}$$

from which

$$\frac{dM_C}{dx} = \frac{Ga}{L} - G_1 \tag{9.4.3}$$

Equation (9.4.3) is the loading position criterion for maximum bending moment at a section of a simple beam; it requires that the value of the expression $(Ga/L - G_1)$ change from positive to negative when a load placed at C is first considered to be at right of C and then at left of C.

Once the critical loading position has been obtained, the value of the maximum bending moment itself may be computed either by using Eq. (9.2.1), or by applying the conventional method for fixed loads.

The position criterion, Eq. (9.4.3), has been derived for the load series as

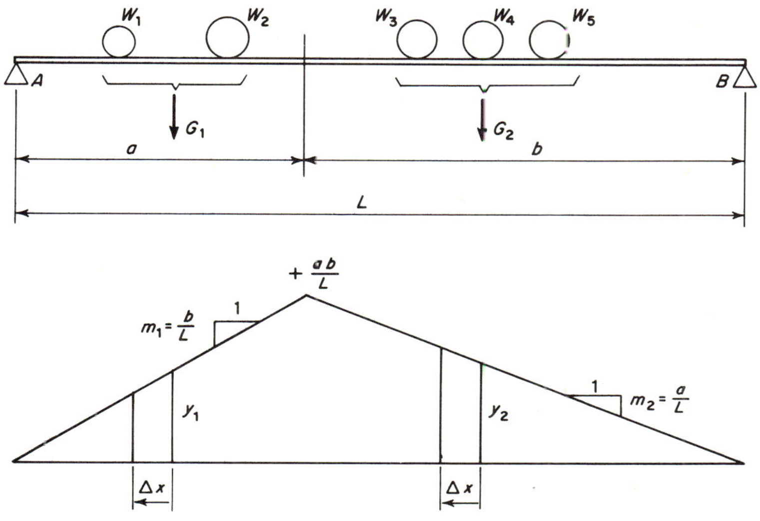

FIG. 9.9. Influence Line for Bending Moment at *C*.

advancing onto the span from the right toward the left. In order to obtain the maximum bending moment at a section *C* due to passage of the load series from left to right, it is only necessary to determine the maximum bending moment at a section *C'* (symmetrical to section *C* with respect to the center line of the beam) due to passage of load from right to left. Employing this method one need not draw another sketch showing the load series in reverse order.

Example 9.3. Determine the maximum bending moment at a section 15 ft from the left support in a 40-ft span simple beam shown in Fig. 9.10a, due to the passage in either direction of the six concentrated loads shown in Fig. 9.10b.

Solution. The position criterion, Eq. (9.4.3), requires that the value of the expression ($Ga/L - G_1$) go from positive to negative when a wheel load placed at the section in question is first excluded from G_1 and then included in G_1. This dictates that the value of Ga/L must lie between the two extreme values of G_1. As indicated by the tabulation:

Wheel at *C*	G_1 on *AC* (excl. wheel at *C*)	$\frac{Ga}{L} = \frac{G(15)}{40} = \frac{3}{8}G$	G_1 on *AC* (incl. wheel at *C*)	Criterion satisfied?
No. 2	5	18.75	13	No
No. 3	13	18.75	28	Yes
No. 4	28	18.75	40	No

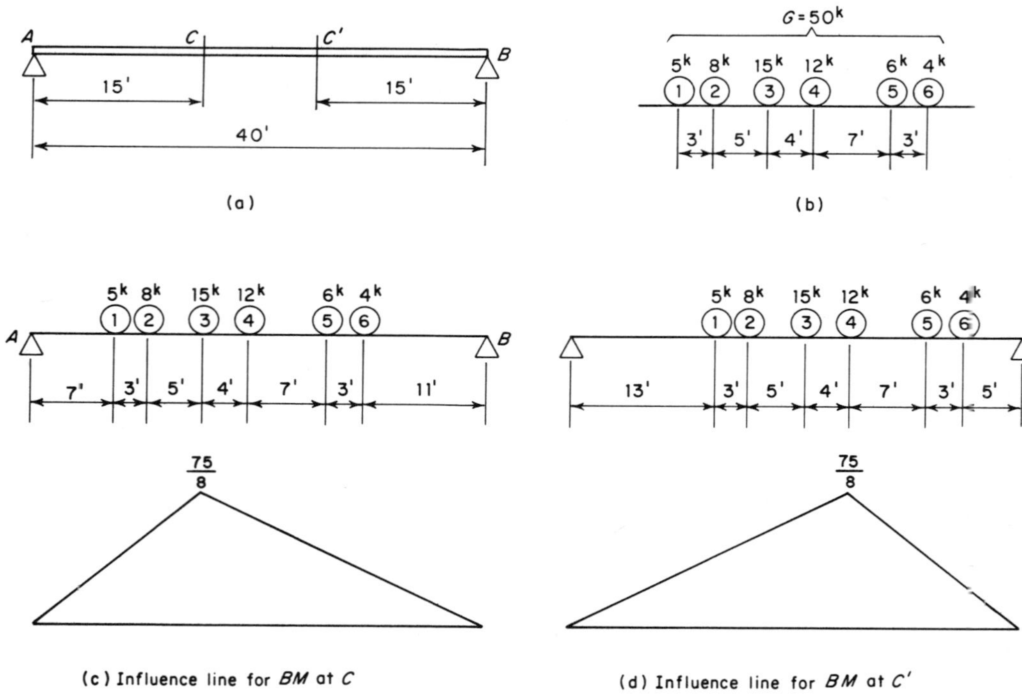

FIG. 9.10. Maximum Bending Moment at a Section of a Simple Beam.

the criterion for maximum M_C is satisfied by placing wheel No. 3 at C. Care must be taken to insure that no wheel load drops off the span in each of the three positions shown in the tabulation. However, if this does occur, each set of G_1 and G values must be consistent by itself. With wheel No. 3 at C, the maximum M_C may be computed by applying Eq. (9.2.1) to the influence line of Fig. 9.10 as

$$\begin{aligned}\text{Max. } M_C &= 5(\tfrac{7}{15})(\tfrac{75}{8}) + 8(\tfrac{10}{15})(\tfrac{75}{8}) + 15(\tfrac{75}{8}) + 12(\tfrac{21}{25})(\tfrac{75}{8}) + 6(\tfrac{14}{25})(\tfrac{75}{8}) + 4(\tfrac{11}{25})(\tfrac{75}{8})\\ &= 21.875 + 50.000 + 140.625 + 94.500 + 31.500 + 16.500\\ &= 355 \text{ ft-kips}\end{aligned}$$

By the conventional method for fixed loads,

$$\begin{aligned}\text{Max. } M_C &= R_A(15) - 5(8) - 8(5)\\ &= \frac{4(11) + 6(14) + 12(21) + 15(25) + 8(30) + 5(33)}{40}(15) - 80\\ &= 29(15) - 80 = 355 \text{ ft-kips (check)}\end{aligned}$$

For maximum bending moment at a section C' 15 ft from the right support, the position criterion, Eq. (9.4.3), is satisfied by placing wheel No. 4 at C', shown by the following tabulation:

Wheel at C'	G_1 on AC' (excl. wheel at C')	$\frac{Ga}{L}=\frac{G(25)}{40}=\frac{5}{8}G$	G_1 on AC' (incl. wheel at C')	Criterion satisfied?
No. 3	13	31.25	28	No
No. 4	28	31.25	40	Yes
No. 5	40	31.25	46	No

Applying Eqs. (9.2.1) to the influence line of Fig. 9.8d,

$$\begin{aligned}\text{Max. } M_{C'} &= 5(\tfrac{13}{25})(\tfrac{75}{8}) + 8(\tfrac{16}{25})(\tfrac{75}{8}) + 15(\tfrac{21}{25})(\tfrac{75}{8}) + 12(\tfrac{75}{8}) + 6(\tfrac{8}{15})(\tfrac{75}{8}) + 4(\tfrac{5}{15})(\tfrac{75}{8}) \\ &= 24.375 + 48.000 + 118.125 + 112.500 + 30.000 + 12.500 \\ &= 345.5 \text{ ft-kips}\end{aligned}$$

By the conventional method for fixed loads,

$$\begin{aligned}\text{Max. } M_{C'} &= R_A(25) - 5(12) - 8(9) - 15(4) \\ &= \frac{4(5) + 6(8) + 12(15) + 15(19) + 8(24) + 5(27)}{40}(25) - 192 \\ &= 21.5(25) - 192 = 345.5 \text{ ft-kips (check)}\end{aligned}$$

Thus the maximum bending at C is 355 ft-kips due to the passage of the load series from right to left; the maximum bending moment at C' is also 355 ft-kips, but it is due to the passage of the load series from left to right.

EXERCISE 9.3

Referring to Fig. 9.11 determine the maximum bending moment at a section 20 ft from the left support in a 50-ft span simple beam due to the passage in either direction of the seven concentrated loads.

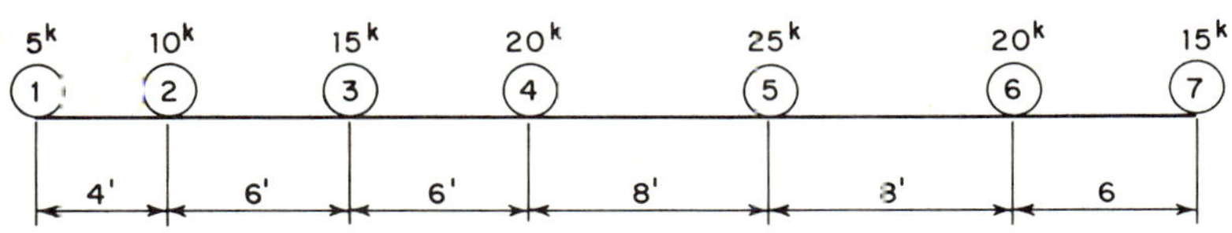

FIG. 9.11. Exercise 9.3.

9.5 Absolute Maximum Bending Moment in a Simple Beam Due to a Long Series of Moving Concentrated Loads

In the preceding article, a method has been presented for obtaining the maximum bending moment at a chosen section of a simple beam due to a long series of moving concentrated loads. It is conceivable that the maximum bending moment thus determined for a section at midspan would be very close to the absolute maximum bending moment which might possibly occur in the beam due to the moving load series. It can be shown, however, that the exact absolute maximum bending moment should occur *at a wheel load* when that wheel load and the resultant force of the load series are at equal distances from the midspan of the beam.

Consider the problem of finding the position of wheel No. 3 in Fig. 9.12 so that the bending moment under it is maximum. From Fig. 9.12, this moment is

$$M = \frac{G(L - s - x)}{L}x - \text{(moment of wheels 1 and 2 about wheel 3)} \tag{9.5.1}$$

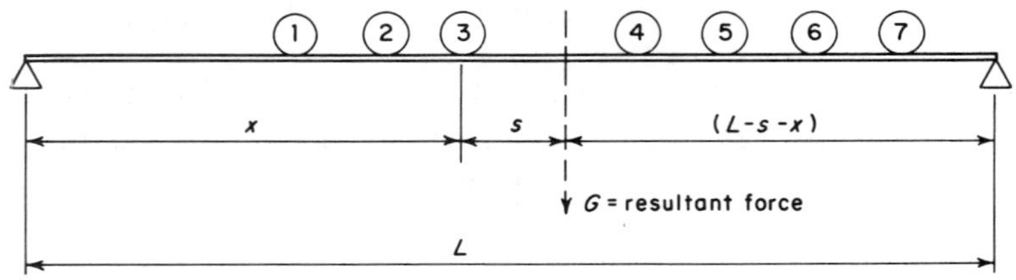

FIG. 9.12. Absolute Maximum Bending Moment in a Simple Beam.

Differentiating Eq. (9.5.1) with respect to x,

$$\frac{dM}{dx} = \frac{G}{L}(L - s - 2x) \tag{9.5.2}$$

Setting Eq. (9.5.2) to zero,

$$x = \frac{L - s}{2} \tag{9.5.3}$$

which makes

$$(L - s - x) = \frac{L - s}{2} \tag{9.5.4}$$

Equations (9.5.3) and (9.5.4) show that wheel No. 3 and the resultant force G should be equidistant from the midspan.

To determine the absolute maximum bending moment in a simple beam due to a series of moving concentrated loads, the maximum bending moment at the midspan of that beam should be determined first. Then the absolute maximum bending moment should occur *under the wheel* which had been placed at the midspan to yield maximum bending moment there. Of course, this wheel and the resultant force should take the position defined by Eqs. (9.5.3) or (9.5.4).

Example 9.4. Determine the maximum bending moment at the center of the 40-ft simple beam shown in Fig. 9.13a, as well as the absolute maximum bending moment in the beam due to the passage of the six concentrated loads shown in Fig. 9.13b.

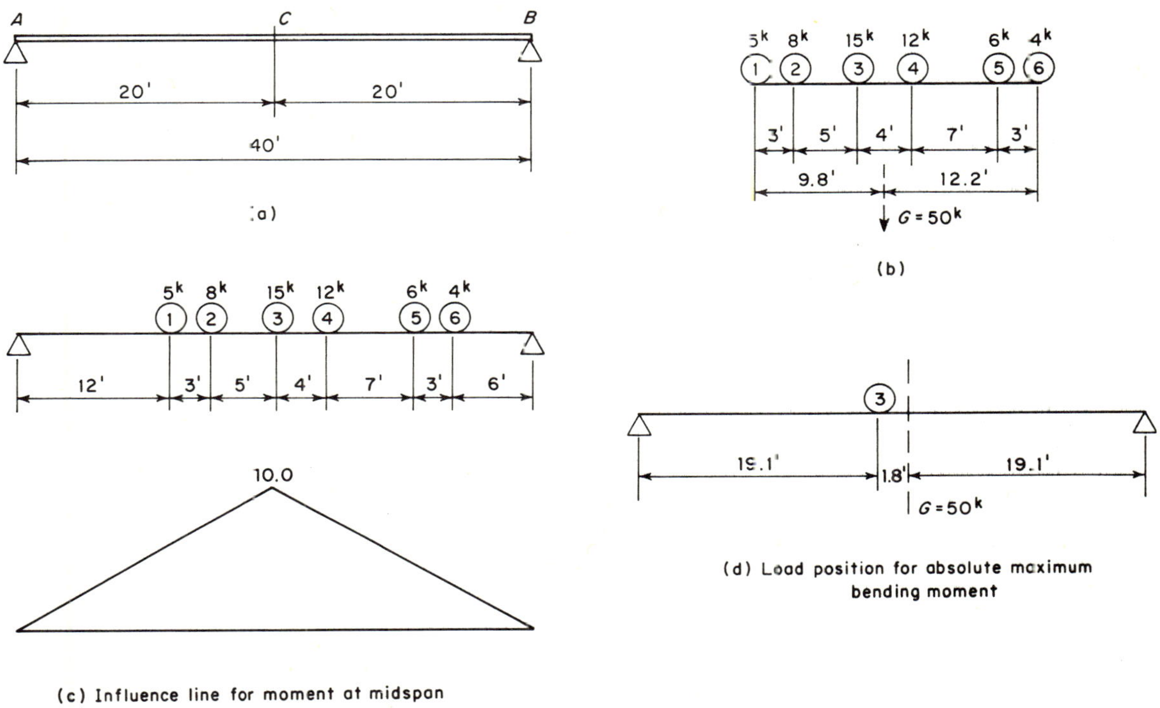

FIG. 9.13. Absolute Maximum Bending Moment in a Simple Beam.

Solution. For maximum bending moment at midspan, the position criterion, Eq. (9.4.3), is satisfied by placing wheel No. 3 at midspan as shown by the following tabulation:

Wheel at C	G_1 on AC (excl. wheel at C)	$\frac{Ga}{L} = \frac{1}{2}G$	G_1 on AC (incl. wheel at C)	Criterion satisfied?
No. 2	5	25	13	No
No. 3	13	25	28	Yes
No. 4	28	25	40	No

Applying Eq. (9.2.1) to the influence line of Fig. 9.13c,

$$\begin{aligned}\text{Max. } M_C &= 5(\tfrac{12}{20})(10) + 8(\tfrac{15}{20})(10) + 15(10) + 12(\tfrac{16}{20})(10) \\ &\quad + 6(\tfrac{9}{20})(10) + 4(\tfrac{6}{20})(10) \\ &= 30 + 60 + 150 + 96 + 27 + 12 = 375 \text{ ft-kips}\end{aligned}$$

By the conventional method for fixed loads,

$$\begin{aligned}\text{Max. } M_C &= R_A(20) - 5(8) - 8(5) \\ &= \frac{4(6) + 6(9) + 12(16) + 15(20) + 8(25) + 5(28)}{40}(20) - 80 \\ &= 455 - 80 = 375 \text{ ft-kips (check)}\end{aligned}$$

From Fig. 9.13b, the resultant force G is at a distance of

$$\frac{6(3) + 12(10) + 15(14) + 8(19) + 5(22)}{50} = 12.2 \text{ ft}$$

from wheel No. 6. The distance s between wheel No. 3 and $G = 50$ kips is 1.8 ft; thus

$$\frac{L - s}{2} = \frac{40 - 18}{2} = 19.1 \text{ ft}$$

From Fig. 9.13d, the absolute maximum bending moment at wheel No. 3 is

$$\begin{aligned}\text{Absolute Max. Moment} &= \frac{50(19.1)^2}{40} - 8(5) - 5(8) \\ &= 456.0125 - 80 = 376.0125 \text{ ft-kips}\end{aligned}$$

It should be noted that the absolute maximum bending moment of 376.0125 ft-kips is only slightly larger than the maximum bending moment of 375 ft-kips at midspan.

EXERCISE 9.4

Determine the maximum bending moment at the center of a 50-ft simple beam, as well as the absolute maximum bending moment in the beam due to the passage of the seven concentrated loads shown in Exercise 9.3 (Fig. 9.11).

9.6 Influence Lines for Reactions on Statically Indeterminate Beams

The influence lines for statically indeterminate structures, unlike those of statically determinate structures, are not composed of straight segments, because the value of the function cannot be easily obtained by the laws of statics as a linear variation with respect to the position along the span of the moving single concentrated load. It will be necessary either to derive an equation of the function for each continuous segment by any convenient method of statically indeterminate structural analysis, or to compute the influence ordinates at as many points along the span as are necessary to yield a smooth influence curve.

As a typical illustration, the influence line for R_B on the two-span continuous beam of Fig. 9.14a is shown in Fig. 9.14d. By the force method of analysis,

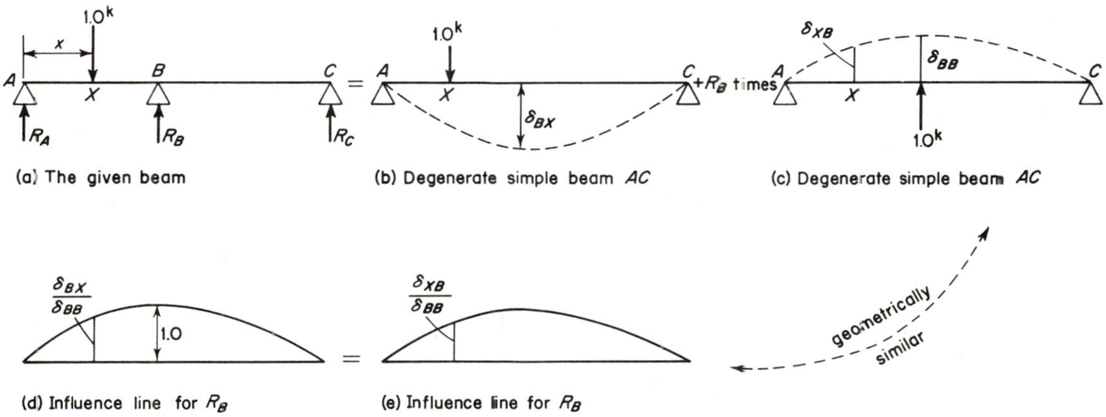

FIG. 9.14. Influence Line for Reaction on a Two-Span Continuous Beam.

for a single concentrated load applied at a distance x from the left support, the compatibility equation may be observed from Fig. 9.11a, b and c to be

$$\delta_{BX} = R_B \delta_{BB}$$

from which

$$R_B = \frac{\delta_{BX}}{\delta_{BB}} \tag{9.6.1}$$

This influence ordinate given by Eq. (9.6.1) is plotted under the single concentrated load in the influence line for R_B in Fig. 9.14d. However, by the Law of Reciprocal Deflections as first discussed in Example 7.4 and later substantiated in general in Art. 8.6,

$$\delta_{BX} = \delta_{XB} \tag{9.6.2}$$

Substituting Eq. (9.6.2) in Eq. (9.6.1) gives the influence ordinate for R_B as

$$R_B = \frac{\delta_{XB}}{\delta_{BB}} \tag{9.6.3}$$

which is shown in Fig. 9.14e. Comparison of Fig. 9.14e with Fig. 9.14c shows that these figures are geometrically similar except that all ordinates on the elastic curve of Fig. 9.14c are to be divided through by the quantity δ_{BB} in order to arrive at the ordinates on the influence line of Fig. 9.14e.

From the preceding discussion it may be observed that the influence line for a reaction on a beam may be obtained: 1) by removing the support for that reaction; and 2) by applying an arbitrary force in the direction of that reaction to yield a unit deflection. In Fig. 9.15, the influence line for R_A is obtained by

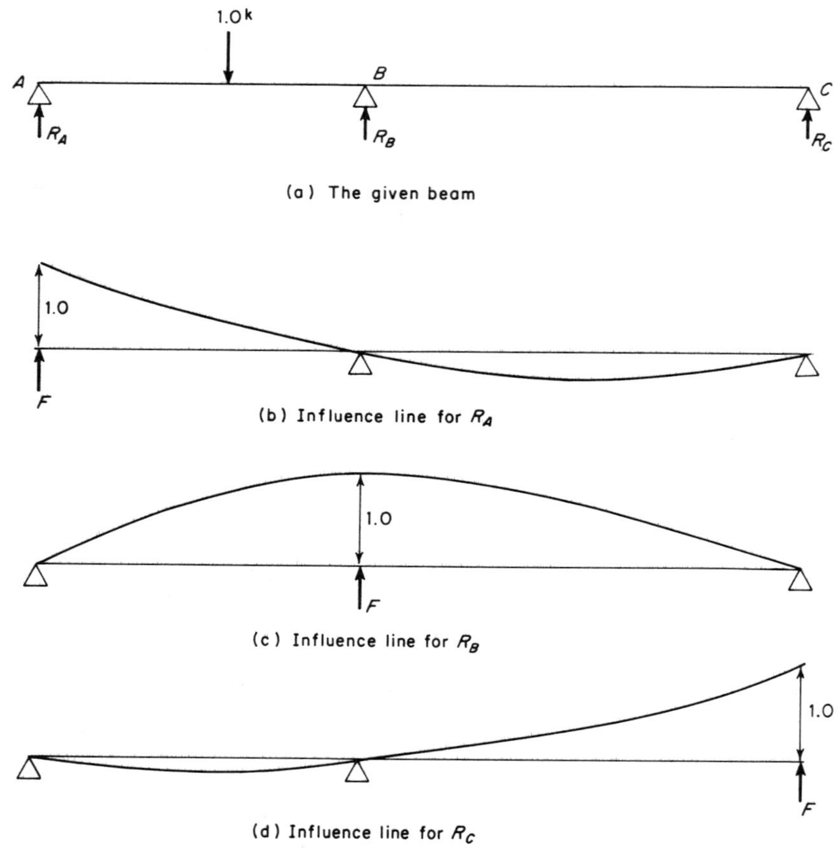

FIG. 9.15. Elastic Curves versus Influence Lines.

applying an arbitrary force at A to the overhanging beam with supports at B and C; for R_B, an arbitrary force at B is applied to the simple beam with supports at A and C; and for R_C, an arbitrary force at C is applied to the overhanging beam with supports at A and B. In fact this same concept may be applied with equal success to the statically determinate beam of Fig. 9.16a,

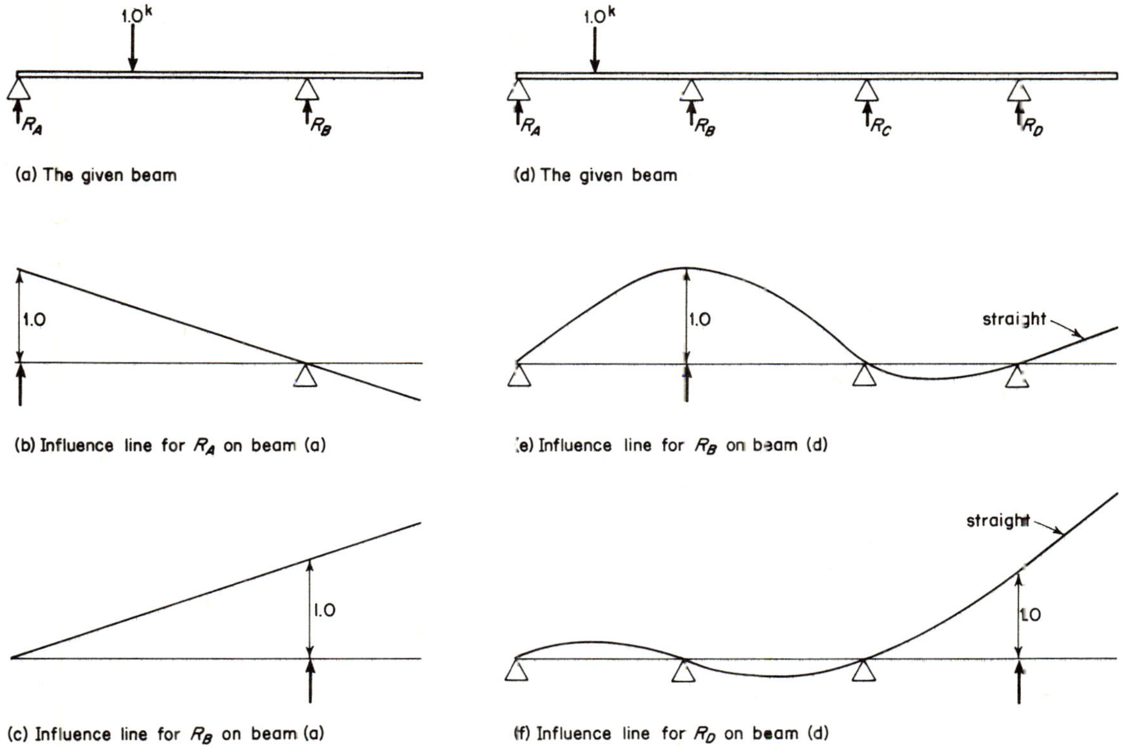

FIG. 9.16. Influence Lines versus Displacement Diagrams.

or to the indeterminate beam of Fig. 9.16d which has more than one degree of indeterminacy. A still more general principle applicable to influence lines for any function (reaction, shear, or bending moment in statically determinate or indeterminate beams or rigid frames, or axial forces in statically determinate or indeterminate trusses, or any other kind of internal force in any kind of complex structures) will be stated in the next article.

Example 9.5. Construct the influence line for the reaction R_A on the statically indeterminate beam of Fig. 9.17a by computing the influence ordinates at 10-ft intervals, noting that these influence ordinates are the deflections on the elastic curve obtained by applying an arbitrary force at A to the overhanging beam with supports at B and C to yield a unit deflection at A.

Solution. Due to the action of a 1-kip arbitrary force at A, the reactions, moment diagram, and elastic curve of the overhanging beam with supports at B and C are shown in Fig. 9.17b, c and d. Applying Conjugate-Beam Theorem I to the elastic curve between B and C,

$$EI\theta_B = \tfrac{2}{3}(M\text{-area on } BC) = \tfrac{2}{3}(750) = 500 \text{ k-ft}^2 \text{ clockwise}$$

$$EI\theta_C = \tfrac{1}{3}(M\text{-area on } BC) = \tfrac{1}{3}(750) = 250 \text{ k-ft}^2 \text{ counterclockwise}$$

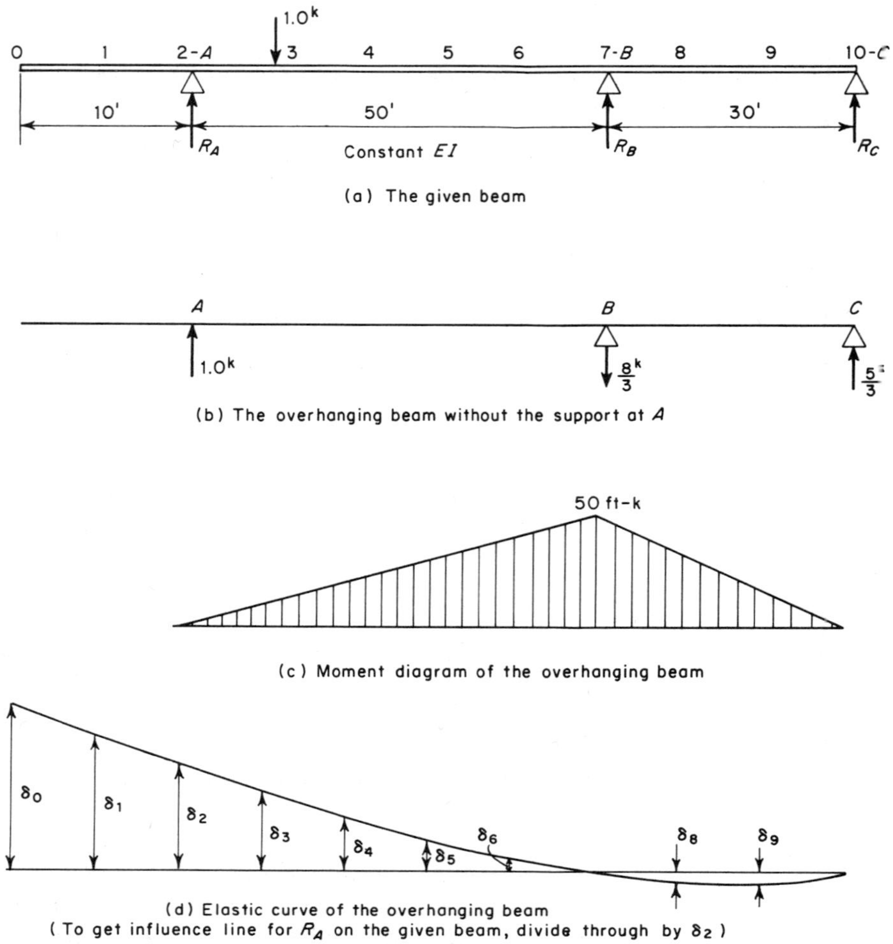

FIG. 9.17. Influence Line for Reaction at *A*.

Applying Conjugate-Beam Theorem II to the elastic curve between *B* and *C*,

$$\begin{aligned} EI\delta_9 &= (EI\theta_C)(10) - (\text{moment of } M\text{-area on } 9\text{-}C \text{ about } 9) \\ &= 250(10) - \tfrac{1}{2}(\tfrac{50}{3})(10)(\tfrac{10}{3}) = 2222 \text{ k-ft}^3 \text{ downward} \\ EI\delta_8 &= (EI\theta_C)(20) - (\text{moment of } M\text{-area on } 8\text{-}C \text{ about } 8) \\ &= 250(20) - \tfrac{1}{2}(\tfrac{100}{3})(20)(\tfrac{20}{3}) = 2778 \text{ k-ft}^3 \text{ downward} \end{aligned}$$

Applying Moment-Area Theorem II to the elastic curve between points 6 and *B*,

$$\begin{aligned} EI\delta_6 &= 10(EI\theta_B) + (\text{moment of } M\text{-area on } 6\text{-}B \text{ about } 6) \\ &= 10(500) + \tfrac{1}{2}(50)(10)(\tfrac{20}{3}) + \tfrac{1}{2}(40)(10)(\tfrac{10}{3}) \\ &= 7333 \text{ k-ft}^3 \text{ upward} \end{aligned}$$

Similarly,

$$EI\delta_5 = 18667 \text{ k-ft}^3 \text{ upward}$$
$$EI\delta_4 = 33000 \text{ k-ft}^3 \text{ upward}$$
$$EI\delta_3 = 49333 \text{ k-ft}^3 \text{ upward}$$
$$EI\delta_2 = 66667 \text{ k-ft}^3 \text{ upward}$$

Applying Moment-Area Theorem I between A and B,

$$EI\theta_A = EI\theta_B + (M\text{-area on } AB)$$
$$= 500 + \tfrac{1}{2}(50)(50) = 1750 \text{ k-ft}^2 \text{ clockwise}$$

From Fig. 9.17d,

$$EI\delta_1 = EI\delta_2 + 10(EI\theta_A) = 66667 + 10(1750)$$
$$= 84167 \text{ k-ft}^3 \text{ upward}$$
$$EI\delta_0 = EI\delta_2 + 20(EI\theta_A) = 66667 + 20(1750)$$
$$= 101667 \text{ k-ft}^3 \text{ upward}$$

The influence ordinates at points 0 to 10 are obtained by dividing δ_0 to δ_{10} by the value of $\delta_2 = 6667$ k-ft^3; they are

$y_0 = +1.5250$	$y_1 = +1.2625$	$y_2 = -1.0000$	$y_3 = +0.7400$
$y_4 = +0.4950$	$y_5 = +0.2800$	$y_6 = -0.1100$	$y_7 = 0.$
$y_8 = -0.0417$	$y_9 = -0.0333$	$y_{10} = 0.$	

EXERCISE 9.5

Construct the influence line for the reaction R_B on the statically indeterminate beam of Example 9.5 by computing the influence ordinates at 10-ft intervals without using the answers to either Example 9.5 or Exercise 9.6.

EXERCISE 9.6

Construct the influence line for the reaction R_C on the statically indeterminate beam of Example 9.5 by computing the influence ordinates at 10-ft intervals without using the answers to either Example 9.5 or Exercise 9.5.

9.7 The Müller-Breslau Influence Theorem

The Müller–Breslau* Influence Theorem may be stated as follows: "The value of a function due to a single concentrated unit load is equal to the deflection

History of Strength of Materials, by S. Timoshenko, McGraw-Hill Book Co., 1953, p. 310.

in the direction opposite to that of the single load at the location of the load, when an arbitrary force or pair of forces is applied at the location of the function on a degenerate structure, obtained by removing the resistance offered by the function, to yield a unit deflection in the direction of the function." This statement seems lengthy; but when applied to the six cases shown in Figs. 9.18 and 9.19, the theorem is simply

$$X = Y \tag{9.7.1}$$

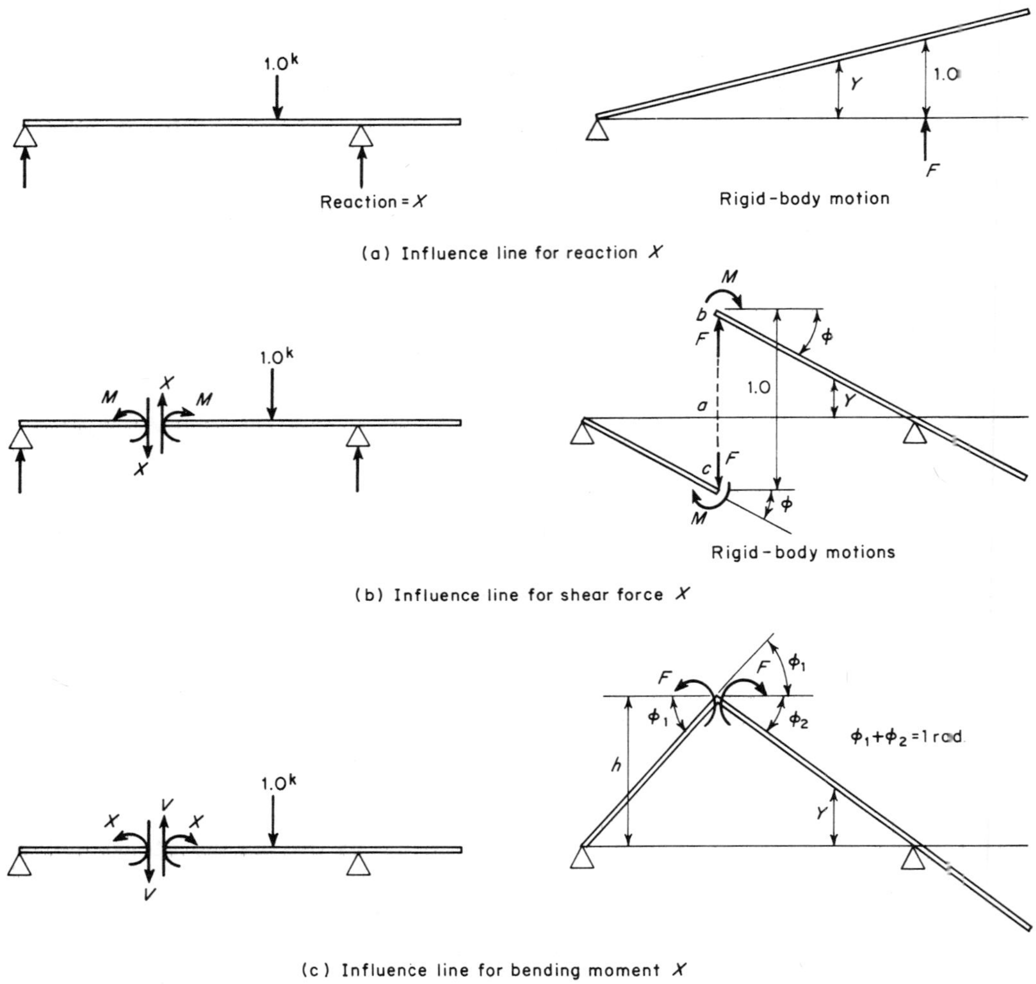

FIG. 9.18. Müller-Breslau Influence Theorem Applied to Statically Determinate Beams.

For statically determinate structures, the proof for Eq. (9.7.1) may be effected by reasoning that the work done by a zero resultant force acting on a rigid body in going through a rigid-body motion of that body has to be zero.

In Fig. 9.18a, the two reactions and the unit-load are in equilibrium; the work done by all three of them in going through the rigid-body motion of

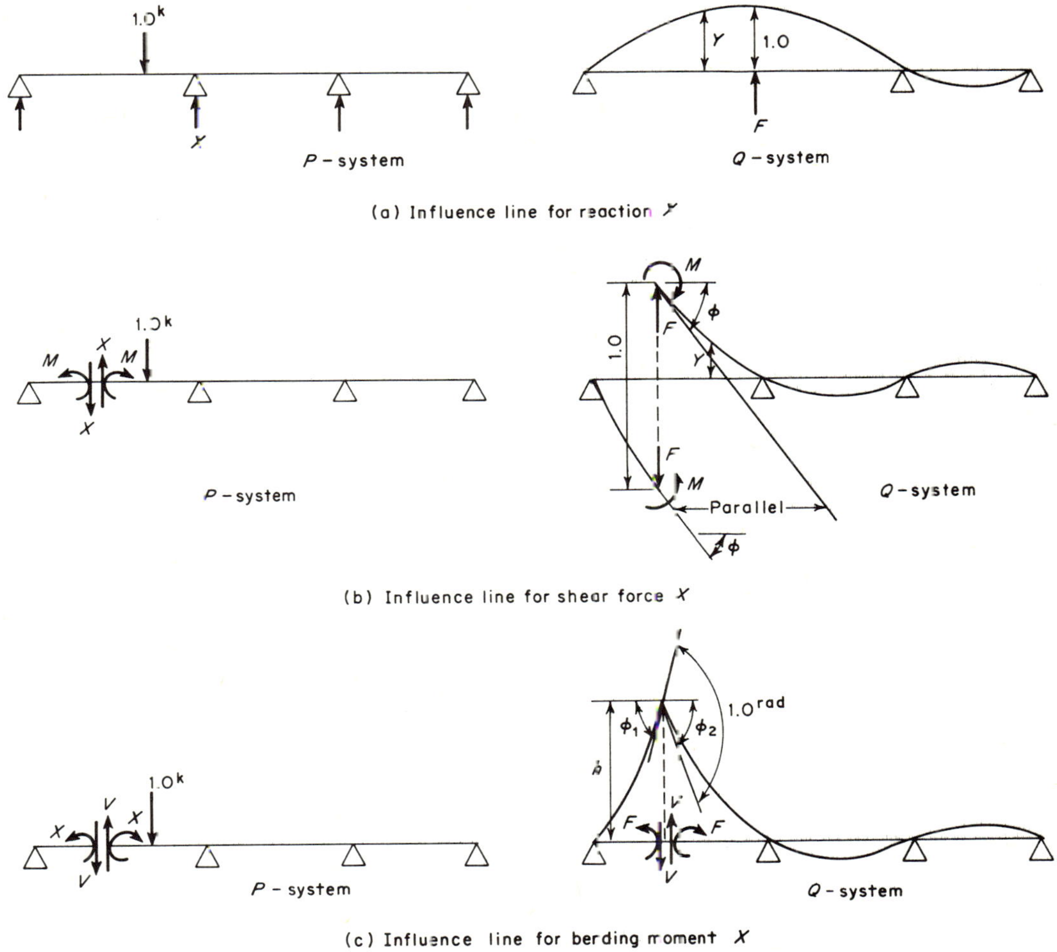

FIG. 9.19. Müller-Breslau Influence Theorem Applied to Statically Indeterminate Beams.

the entire beam is

$$0 = -(1.0\ \text{kip})(Y\ \text{in.}) + (X\ \text{kips})(1.0\ \text{in.})$$

In. Fig. 9.18b, there are two sets of forces in equilibrium and two rigid-body motions; the work done by the first set is

$$0 = +(X)(ac) - (M)(\phi)$$

and the work done by the second set is

$$0 = +(X)(ab) + (M)(\phi) - (1.0)(Y)$$

The sum of the works done by both sets is still zero; or

$$0 = +X(ab + ac) - (1.0)(Y) = +(X)(1.0) - (1.0)(Y)$$

Note that the two rigid bodies are parallel because there is still continuity in slope maintained by the moment resistance at the section.

In Fig. 9.18c, there are again two sets of forces in equilibrium and two rigid-body motions; the work done by the first set is

$$0 = +(X)(\phi_1) - (V)(h)$$

and the work done by the second set is

$$0 = +(X)(\phi_2) + (V_1)(h) - (1.0)(Y)$$

The sum of the work done by both sets is still zero; or

$$0 = +(X)(\phi_1 + \phi_2) - (1.0)(Y) = +(X)(1.0) - (1.0)(Y)$$

For statically indeterminate structures, the proof for Eq. (9.7.1) may be made by applying the Reciprocal Energy Theorem substantiated in Art. 8.6: the work done by a *P*-force system, in going through the displacements of a *Q*-force system, is equal to the work done by the *Q*-force system in going through the displacements of the *P*-force system. In applying this reciprocal theorem to Fig. 9.19, one must keep in mind that either the *P*-force system or the *Q*-force system is to be applied to the degenerate structure without the physical restraint associated with the specific function.

In Fig. 9.19a,

$$P * \Delta Q = -(1.0)(Y) + (X)(1.0)$$

but,

$$Q * \Delta P = 0$$

because there is no displacement in the *P*-system for the arbitrary force *F* in the *Q*-system.

In Fig. 9.19b,

$$\begin{aligned} P * \Delta Q &= -(M)(\phi) + (M)(\phi) + (X)(ab) + X(ac) - (1.0)(Y) \\ &= +(X)(1.0) - (1.0)(Y) \end{aligned}$$

but,

$$Q * \Delta P = 0$$

because there is continuity in both slope and deflection at the section being considered in the *P*-system.

In Fig. 9.19c,

$$\begin{aligned} P * \Delta Q &= -(V)(h) + (V)(h) + (X)(\phi_1) + (X)(\phi_2) - (1.0)(Y) \\ &= +(X)(1.0) - (1.0)(Y) \end{aligned}$$

but,

$$Q * \Delta P = 0$$

again because there is continuity in both slope and deflection at the section being considered in the P-system.

The Müller–Breslau Influence Theorem is useful as an aid in making free-hand sketches of influence lines in a qualitative manner, and in actually producing quantitative influence lines by means of models in the laboratory.

9.8 Concluding Remarks

The subject of influence lines is a profound one, and its treatment in this chapter has been limited to a few basic concepts. In the days when truss or girder bridges were built along railroads to accommodate the passages of double locomotives in either direction, some complex and interesting analysis problems were encountered. Because of the smaller number of concentrated wheel loads, the critical loading positions of highway truck loading systems can usually be ascertained by visual inspection, thus lessening the need of derivation for many algebraic criteria in different situations. In the case of statically indeterminate structures, so long as the influence lines are composed of curved segments, the critical position of a concentrated load series has to be determined by trial and error.

Index

HIGH PERFORMANCE COMPUTING AND COMMUNICATIONS

HEARING

BEFORE THE

SUBCOMMITTEE ON SCIENCE

OF THE

COMMITTEE ON SCIENCE, SPACE, AND TECHNOLOGY

U.S. HOUSE OF REPRESENTATIVES

ONE HUNDRED THIRD CONGRESS

SECOND SESSION

MAY 10, 1994

[No. 148]

Printed for the use of the
Committee on Science, Space, and Technology

U.S. GOVERNMENT PRINTING OFFICE
84–115 CC WASHINGTON : 1994

For sale by the U.S. Government Printing Office
Superintendent of Documents, Congressional Sales Office, Washington, DC 20402
ISBN 0-16-046307-6

COMMITTEE ON SCIENCE, SPACE, AND TECHNOLOGY

GEORGE E. BROWN, Jr., California, *Chairman*

MARILYN LLOYD, Tennessee
DAN GLICKMAN, Kansas
HAROLD L. VOLKMER, Missouri
RALPH M. HALL, Texas
DAVE McCURDY, Oklahoma
TIM VALENTINE, North Carolina
ROBERT G. TORRICELLI, New Jersey
RICK BOUCHER, Virginia
JAMES A. TRAFICANT, Jr., Ohio
JAMES A. HAYES, Louisiana
JOHN S. TANNER, Tennessee
PETE GEREN, Texas
JIM BACCHUS, Florida
TIM ROEMER, Indiana
ROBERT E. (BUD) CRAMER, Jr., Alabama
DICK SWETT, New Hampshire
JAMES A. BARCIA, Michigan
HERBERT C. KLEIN, New Jersey
ERIC FINGERHUT, Ohio
PAUL McHALE, Pennsylvania
JANE HARMAN, California
DON JOHNSON, Georgia
SAM COPPERSMITH, Arizona
ANNA G. ESHOO, California
JAY INSLEE, Washington
EDDIE BERNICE JOHNSON, Texas
DAVID MINGE, Minnesota
NATHAN DEAL, Georgia
ROBERT C. SCOTT, Virginia
XAVIER BECERRA, California
PETER W. BARCA, Wisconsin
BOBBY L. RUSH, Illinois

ROBERT S. WALKER, Pennsylvania*
F. JAMES SENSENBRENNER, Jr., Wisconsin
SHERWOOD L. BOEHLERT, New York
TOM LEWIS, Florida
HARRIS W. FAWELL, Illinois
CONSTANCE A. MORELLA, Maryland
DANA ROHRABACHER, California
STEVEN H. SCHIFF, New Mexico
JOE BARTON, Texas
DICK ZIMMER, New Jersey
SAM JOHNSON, Texas
KEN CALVERT, California
MARTIN R. HOKE, Ohio
NICK SMITH, Michigan
EDWARD R. ROYCE, California
ROD GRAMS, Minnesota
JOHN LINDER, Georgia
PETER BLUTE, Massachusetts
JENNIFER DUNN, Washington
BILL BAKER, California
ROSCOE G. BARTLETT, Maryland
VERNON J. EHLERS, Michigan

ROBERT E. PALMER, *Chief of Staff*
MICHAEL RODEMEYER, *Chief Counsel*
KATHRYN HOLMES, *Administrator*
DAVID D. CLEMENT, *Republican Chief of Staff*

SUBCOMMITTEE ON SCIENCE

RICK BOUCHER, Virginia, *Chairman*

RALPH M. HALL, Texas
TIM VALENTINE, North Carolina
JAMES A. BARCIA, Michigan
DON JOHNSON, Georgia
ANNA G. ESHOO, California
E. B. JOHNSON, Texas
DAVID MINGE, Minnesota
PETER W. BARCA, Wisconsin

SHERWOOD L. BOEHLERT, New York
JOE BARTON, Texas
SAM JOHNSON, Texas
NICK SMITH, Michigan
PETER BLUTE, Massachusettes

*Ranking Republican Member.